D0935213

Solid Pharmaceutics: Mechanical Properties and Rate Phenomena

Solid Pharmaceutics: Mechanical Properties and Rate Phenomena

JENS THURØ CARSTENSEN

Center for Health Sciences
School of Pharmacy
University of Wisconsin
Madison, Wisconsin

1980

ACADEMIC PRESS

A Subsidiary of Harcourt Brace Jovanovich, Publishers

New York London Toronto Sydney San Francisco

ACADEMIC PRESS, INC.
111 Fifth Avenue, New York, New York 10003

United Kingdom Edition published by
ACADEMIC PRESS, INC. (LONDON) LTD.
24/28 Oval Road, London NW1 7DX

Library of Congress Cataloging in Publication Data

Carstensen, Jens Thorø, Date.
Solid Pharmaceutics: Mechanical Properties and Rate Phenomena
Includes bibliographical references and index.
1. Solid dosage forms. 2. Solid dosage forms-- Mechanical properties. I. Title. [DNLM: 1. Dosage forms. 2. Tablets. 3. Powders. 4. Technology, Pharmaceutical. QV785 C329s]
RS201.S57C37 615'.191 79-6805
ISBN 0-12-161150-7

PRINTED IN THE UNITED STATES OF AMERICA

80 81 82 83 9 8 7 6 5 4 3 2 1

To my wife
Winifred Rikkers Carstensen

Contents

Appendix

Preface

Pharmaceutics plays an important role in the field of pharmacy. The integrity of the solid dosage form sold to the consumer is of great concern. And, because the final form and quality of this product are influenced by a multitude of factors, the solid pharmaceuticist must be well versed in many areas of science: solid state physics, physical chemistry, organic chemistry, and engineering science.

Much research has been done on the basic principles underlying this branch of the applied sciences. The book attempts to consolidate important findings in the field of solid pharmaceutics. It reviews basic scientific and technological aspects of pharmaceutics from the molecular properties of single crystals through the properties of single component, multiparticulate systems to binary mixtures. Other important topics, such as the rheology of powders and the physics of powder compression, are also discussed.

This book should serve to acquaint the advanced reader with the field of solid state pharmaceutics, while at the same time it provides the research scientist with updates on a variety of timely topics and, hopefully, with an incentive and guide for future research projects.

I am grateful for the contribution that my wife, Winifred Carstensen, has made by both typing a large portion of the manuscript and doing a considerable number of the drawings. Also, I am most thankful for the work of Dr. Michael Zoglio, Dr. Pierre Toure, and Mr. Ashok Mehta; they have provided much technical help with several chapters. Finally, I am deeply grateful for the encouragement and stimulation that Dean George Zografi and Professor Francis Puisieux offered throughout the duration of the project.

CHAPTER

I

Single-Compound Systems

Most dosage forms are multiple-component systems, but in order to classify and to appreciate the properties of multicomponent systems, it is necessary to deal first with the simpler system consisting of one compound only. Chapters I and II will therefore deal with systems of one compound only.

It is convenient to divide the treatment into properties on the atomic and molecular level and properties of the bulk. The former will be denoted microscopic properties and the latter macroscopic properties, and discussions of both single- and multiple-component systems will treat first microscopic and then macroscopic properties.

I-1 CRYSTAL SYSTEMS

Substances used in solids and solid dosage forms in pharmaceutics are mostly organic compounds and these are usually molecular crystals or amorphic substances. At times salts are used, and these are ionic crystals. We shall first deal with crystalline substances.

A crystalline substance is one in which the molecules or ions are positioned in a lattice. A lattice is a geometric array which is characterized by periodicity.

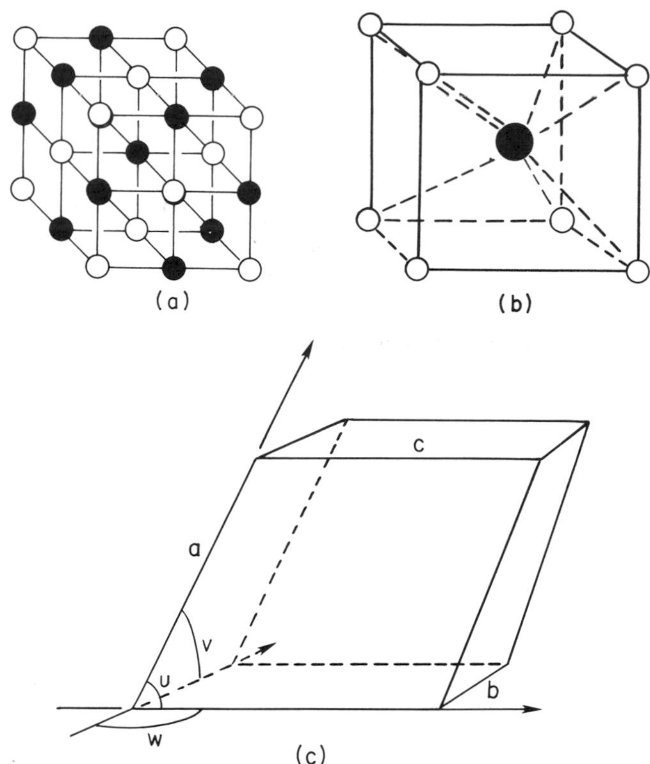

Fig. I-1 (a) Arrangement of 1 : 1 electrolyte in a face centered cubic lattice; (b) 1 : 1 electrolyte in a body centered cubic lattice; (c) triclinic crystal.

Some simple crystal systems are shown in Fig. I-1. For further details on crystal systems the reader is referred to texts on the subject (Mullin, 1961; Kittel, 1956).

In the face centered cubic crystal shown in Fig. I-1a each ion is surrounded (at equal distance) by six different ions. One speaks in this case of a coordination number of 6. An example of this type of cubic crystal is sodium chloride. In the body centered cubic crystal shown in Fig. 1b the coordination number is 8; an example of this is potassium chloride.

To describe a crystal (system) one places it in a coordinate system, as shown, e.g., in Fig. 1c. The lattice constants are a, b, and c (and in the hexagonal system d), and the angles between the axes are u, v, and w.

Depending on angles (whether right or oblique) and lattice constants (whether equal or different) there are a number of combinations that lead to different crystal systems. The seven crystal systems are listed in Table I-1.

TABLE I-1

The Crystal Systems

Angles between axes	Length of axes	System	Alternate system name	Examples
$u = v = w = 90°$	$a = b = c$	Regular	Cubic	NaCl
$u = v = w = 90°$	$a = b \neq c$	Tetragonal	Pyramidal	Rutile
$u = v = w = 90°$	$a \neq b \neq c$	Orthorhombic	Rhombic	Silver nitrate
$u = v = 90° \neq w$	$a \neq b \neq c$	Monoclinic		*p*-aminobenzoic acid
$u \neq v \neq w \neq 90°$	$a \neq b \neq c$	Triclinic		Potassium dichromate
$u = v = w \neq 90°$	$a = b = c$	Trigonal	Rhombohedral	Sodium nitrate
Three 60°, one 90°	$a = b = c \neq d$	Hexagonal		Graphite

I-2 IONIC RADII

The lattice constants are obtained by means of x-ray diffraction. Bragg's law states that

$$n\lambda = 2R_0 \sin \alpha \tag{I-2-1}$$

where n is an integer, λ is the wavelength of the x ray, R_0 is the lattice constant, and α is one of the angles giving rise to attenuation. Note that in ionic crystals one obtains $R_0 = r_+ + r_-$, where r is the ionic radius. This term is but vaguely defined since there is an exponential tailing of the radial wave function, but a formal number may be obtained in a couple of ways and has utility, as will be seen shortly. (a) In a metal halide, for instance, if M denotes a metal and X a halide, r_{M^+} may be assumed to equal that in the metal, and r_{X^-} is then equal to $R_0 - r_{M^+}$. (b) It may be obtained by use of screening constants as outlined below.

In NaF, $R_0 = 2.31$, and the question is how much is r_{Na^+} and how much r_{F^-}. The higher the nuclear charge Z, the more the "pull" on the outermost electrons, and hence the smaller r.

It can be shown (Pitzer, 1953, p. 215) that

$$r/a_0 = n^2/Z' \tag{I-2-2}$$

where a_0 ($= 0.529$ Å) is the radius of the first Bohr orbital, n is the principal quantum number, and Z' is the effective nuclear charge. This latter is the actual nuclear charge minus the screening constant S, which can be calculated quantum mechanically from the noble gas configurations (Pitzer, 1953, p. 215).

If $n_+ = n_-$, for instance, then

$$r_+ = q/Z'_+ \qquad \text{(I-2-3a)}$$

$$r_- = q/Z'_- \qquad \text{(I-2-3b)}$$

$$R = r_+ + r_- = q[(1/Z'_+) + (1/Z'_-)]$$

or

$$q = Z'_+ Z'_- R_0/(Z'_+ + Z'_-) \qquad \text{(I-2-4)}$$

where q is a constant. Inserting the value of q from Eq. (I-2-4) into Eq. (I-2-3a) then gives

$$r_+ = Z'_- R_0/(Z'_+ + Z'_-) \qquad \text{(I-2-5)}$$

For example, the neon structure, applicable to Na^+ and F^-, has a value of $S = 4.52$, so

$$Z'_{Na^+} = 11 - 4.52 = 6.48$$

$$Z'_{F^-} = 9 - 4.52 = 4.48$$

i.e.,

$$r_{Na^+} = [4.48/(6.48 + 4.48)]2.31 = 0.95$$

This compares favorably with the accepted value shown in Table I-2.

For ionic compounds the hardness frequently increases in a series with decreasing interionic distance. Melting points frequently decrease in a particular series with increasing interionic distance, but with relation to melting point distances are obviously only one factor. The surface energy of crystals (at 0°K) decreases within a series with increasing interionic distance (Shuttleworth, 1949).

One utility of the ionic radii is the qualitative value of what is known as the radius ratio rule. Consider, for instance, a binary ionic compound with coordination number 3 in the configuration shown in Fig. I-2. In the nomenclature of Fig. I-2b line AE has the equation

$$y = (\sqrt{3}/3)x \qquad \text{(I-2-6)}$$

TABLE I-2

Ionic Radii of Alkali Metals and Halides

Substance	Radius (Å)	Substance	Radius (Å)
Li^+	0.68	F^-	1.33
Na^+	0.97	Cl^-	1.81
K^+	1.33	Br^-	1.96
Rb^+	1.47	I^-	2.20
Cs^+	1.67		

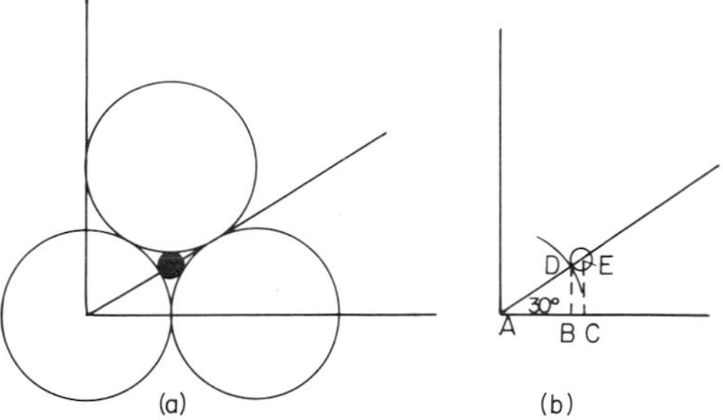

Fig. I-2 Binary ionic compound with ions of different radius (a) showing small-ion fitting in interstitial space of larger ions and (b) showing the geometry for calculating the radius ratio.

Point D is the intersection of this line and the circle

$$y^2 + x^2 = r^2 \tag{I-2-7}$$

Hence the intersection abscissa is

$$x_1^2/3 + x_1^2 = r^2 \tag{I-2-8}$$

so that

$$x_1 = (\sqrt{3}/2)r \tag{I-2-9a}$$

and

$$y_1 = r/2 \tag{I-2-9b}$$

The coordinates of point E are given by

$$x_2 = r \tag{I-2-10a}$$

so that

$$y_2 = (\sqrt{3}/3)r \tag{I-2-10b}$$

so that the distance ED is given by

$$\text{ED} = r\left\{\left(1 - \sqrt{3}/2\right)^2 + \left[\left(\sqrt{3}/3\right) - \tfrac{1}{2}\right]^2\right\}^{1/2} = 0.155r \tag{I-2-11}$$

A similar calculation for a tetrahedron will show that ED here is $0.225r$; hence between radius ratios of 0.155 and 0.225 a coordination number of 3 will be geometrically dictated. Table I-3 shows radius ratios of various lattices.

It is seen, therefore, that in general the geometry dictates the crystal system of a salt. It is also the size more than the charge which dictates substitution of one ion for another in a lattice. This will be dealt with more

TABLE I-3

Radius Ratios and Coordination Numbers

r_A/r_B	Coordination number	Lattice	Example
0–0.155	2		Carbon dioxide
0.155–0.225	3	Hexagonal	Boron nitride
0.225–0.414	4	Tetrahedral	Zinc blende
0.414–0.733	6	Octahedral	Sodium chloride
0.733–1.0	8	Body centered cubic	Cesium chloride
1.0	12	Face centered cubic Hexagonal	

fully under defects, but as examples Na^+ (0.95 Å) will substitute for Ca^{++} (0.99 Å), Li^+ (0.60 Å) will substitute for Mg^{++} (0.65 Å) but not for Na^+ (0.95 Å), and Mg^{++} (0.65 Å) will substitute for Fe^{++} (0.76 Å) and Fe^{+++} (0.64 Å) but not for Ca^{++} (0.99 Å).

I-3 LATTICE ENERGIES OF IONIC CRYSTALS

By considering each ion in an ionic crystal subject to an electostatic attraction the energy of which is

$$-Z_1Z_2e^2/r \tag{I-3-1}$$

and a repulsion of an energy of

$$\alpha/r^n \tag{I-3-2}$$

where Ze is ionic charge, r is ionic separation, α is a constant, and n is a number between 6 and 12, one can show that for a crystal with $Z_1 = Z_2 = 1$ and $2N$ ions, the crystal energy U is given by (Fig. I-3a)

$$U = (NAe^2/R_0)(1 - 1/n) \tag{I-3-3}$$

where A is the so-called Madelung constant and is given by

$$A = \sum(\pm)p_{ij}^{-1} \tag{I-3-4}$$

where i and j are running indices that run from 1 to N but are never equal. The Madelung constant A can be obtained by systematic summation, and, e.g., for sodium chloride has the value 1.75.

The crystal energy U can be obtained from a Born–Haber cycle (Kittel, 1956, p. 9), and hence by equating this with Eq. (I-3-3) and inserting A and R_0, n can be evaluated. For NaCl, e.g., it is 8.

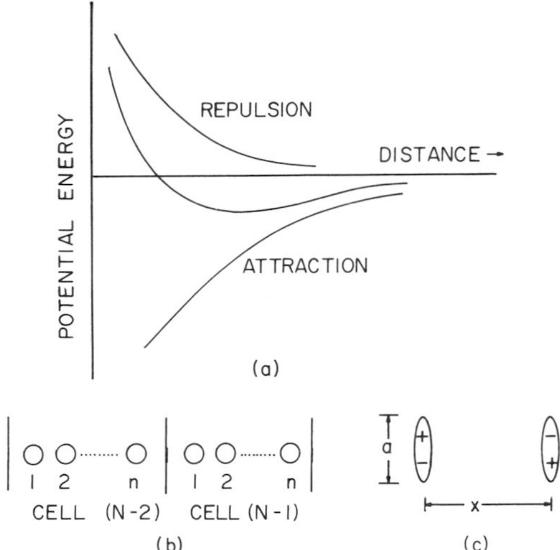

Fig. I-3 (a) Potential energy diagram in a crystal as a function of the distance between units making up the crystal. (b) Cell arrangement for use in calculations for organic molecular crystals (Walmsley, 1968). (c) Arrangement of dipoles for dipole interactions in crystals.

Furthermore (Slater, 1924), since the compressibility of a substance is given by K in the expression

$$K = -(1/V)(\partial V/\partial P)_T \tag{I-3-5}$$

and since $dU = -p\,dV$ and $V = 2NR_0^3$, it can be shown that

$$1/K = (n-1)e^2A/18R_0^2 \tag{I-3-6}$$

From experimental compressibility data, one can then determine n, which by this method correlates well with that found using the Born–Haber value in the theoretical treatment of Eq. (I-3-3). The details of the above calculations may be found in appropriate texts [e.g., Carstensen (1973)].

I-4 LATTICE ENERGIES OF MOLECULAR CRYSTALS

In the case of molecular crystals (as opposed to ionic, covalent, or metallic crystals) one encounters low heats of sublimation. This is also in contrast to the much greater dissociation energies required to separate an organic molecule into its constituent atoms. In one approach to obtaining lattice energies (Born and Huang, 1954) the lattice is divided into cells each containing one lattice period (Fig. I-3b being a presentation of this in one dimension). The positional characteristic of an atom is then the vector

$$x\begin{pmatrix} j \\ k \end{pmatrix} \tag{I-4-1}$$

so that a displacement from equilibrium position x_0 is

$$u\begin{pmatrix} j \\ k \end{pmatrix} = x\begin{pmatrix} j \\ k \end{pmatrix} - x_0\begin{pmatrix} j \\ k \end{pmatrix} \tag{I-4-2}$$

The potential energy ϕ, is then expanded about the equilibrium potential energy ϕ_0 and coordinates by a Taylor series. Walmsley (1968) has developed an expression for lattice vibration frequencies and elastic stiffness constants for molecular crystals using a molecular (rather than atomic) model and assuming that the molecules are rigid and that only the sum of the interactions of pairs of molecules contribute to the potential energy. For carbon dioxide at 0°K an average elastic stiffness coefficient of 9.4×10^{10} dyn/cm^2 was calculated from the theory. The experimental values at 145 and 77°K extrapolate to 10×10^{10} dyn/cm^2 at 0°K.

In general, the repulsive term arises because of orbital overlaps (Pellegrini *et al.*, 1973). These authors calculated the lattice energy for a simple hydrogen-bonded molecule like methanol. The methanol is present in the lattice structure as dimers. The minimum potential energy ϕ_L was calculated as the sum of the interaction energy ϕ_{AB}^{HB} between two hydrogen-bonded molecules A and B and the energy ϕ_{AK} associated with all the interactions of *one* molecule A with the surrounding molecules, i.e.,

$$\phi_L = \phi_{AB}^{HB} + \phi_{AK} \tag{I-4-3}$$

and typically ϕ_L was found to be $-5 - 6 = -11$ kcal/mole. The manner in which ϕ_{AK} was found was the method of Ferro and Hermans (1972) where

$$\phi_{AK} = \sum_{ij} \left(B_{ij} R_{ijk}^{-12} - A_{ij} R_{ijk}^{-6} + C_{ij} R_{ijk}^{-1} \right) \tag{I-4-4}$$

The Coulombic coefficient C_{ij} is proportional to the product of the point charges on molecules i and j, assuming that the dipole moment is the same as the dipole moment in the vapor phase. The coefficients A and B are nonbonded coefficients and ϕ_{AB}^{HB} is computed based on a Hartree–Fock model.

In certain instances the arrangement of the organic molecules in the lattice is comparable to dipoles stacked in alternating directions (Fig. I-3c). In this case the dipolar charges (whether actual or induced dipoles) are x cm apart so that the dipole moments are $\mu = ae$. The force is then

$$\left[-2e^2/(x^2 + a^2) \right] + 2e^2/x^2 = 2\mu^2/x^4 \tag{I-4-5}$$

and the energy is $2\mu^2/3x^3$, so that if we set the central field force proportional to x^{-n} the potential energy may be written

$$\phi = (\lambda/p_{ij}^n x^n) \pm 2\mu^2/p_{ij}^3 3x^3 = (\lambda A/x^n) - 2\mu^2\alpha/3x^3 \qquad \text{(I-4-6)}$$

where A and α (in a fashion similar to the treatment of ionic crystals) are

$$A = \sum_{ij}{}' p_{ij}^{-n} \qquad \text{and} \qquad \alpha = -\sum_{j}{}' \pm p_{ij}^{-3} \qquad \text{(I-4-7)}$$

At equilibrium $\partial\phi/\partial x = 0$, so that

$$A\lambda/x^n = 2\alpha\mu^2/nx^3 \qquad \text{(I-4-8)}$$

which inserted in the expression for ϕ above and considering that the energy is N gives

$$U = N\phi = (N\alpha\mu^2/x^3)\left[(2/n) - \tfrac{2}{3}\right] \qquad \text{(I-4-9)}$$

For urea, for example, $\mu = 5 \times 10^{-18}$ esu and the coordination number is 6; $x = 5 \times 10^{-8}$ cm and if one assumes that $\alpha = 1$ (i.e., that $2^{-3} + 3^{-3} + \cdots = 0$), then the energy is calculated to be

$$U = 6(6 \times 10^{23})(5 \times 10^{-18})^2 \tfrac{1}{3} / (5 \times 10^{-8})^3$$
$$= 3 \times 10^{11} \text{ erg/mole} = 40 \text{ kJ/mole} = 10 \text{ kcal/mole} \qquad \text{(I-4-10)}$$

in good agreement with experimental values.

Lattice energies of organic molecules are frequently obtained experimentally [as suggested by Davies (1959, 1971), Davies and Jones (1959), Davies and Kybett (1963, 1965), Davies and Malpass (1961), and Davies *et al.* (1959)] by determining vapor pressures; the ensuing Clausius–Clapeyron plots have slopes equal to the heats of sublimation. This must be extrapolated to 0°K (since the heat contents of the molecules in the gas phase must be adjusted for). Experimental values of the lattice energy obtained in this fashion have been reported for a series of compounds.

I-5 LATTICE VIBRATIONS IN CRYSTALS

Figure I-4a depicts the equilibrium positions (circles) of (identical) molecules in a one-dimension crystal. The lines denote (hypothetical) actual positions of the molecules at some time t, and the arrows indicate the directions of the forces.

If only nearest neighbors interact, then the force on the nth molecule will be f, which is given by

$$f = B(u_{n+1} - u_n) - B(u_n - u_{n-1}) = B(u_{n+1} + u_{n+2} - 2u_n) \qquad \text{(I-5-1)}$$

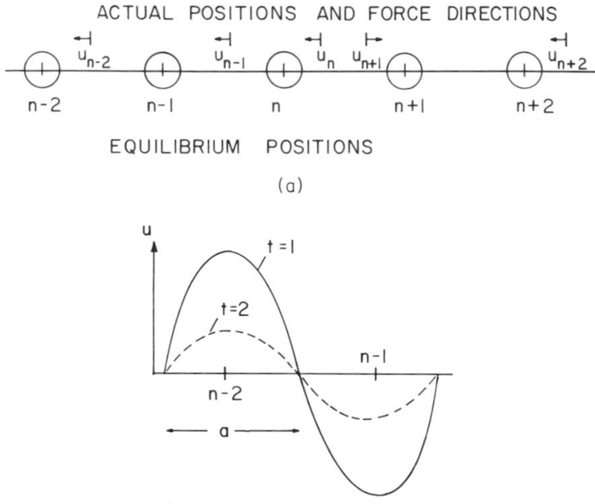

Fig. I-4 (a) Positional coordinates in a two-dimensional crystal. (b) Deviations u from the equilibrium position in the standing wave situation.

Here B is a force constant. The force equals the mass M times the acceleration $\ddot{u}$ so that

$$M\ddot{u}_n = B(u_{n+1} + u_{n+2} - 2u_n) \tag{I-5-2}$$

One seeks solutions of the type

$$u_n = A \exp[i(\omega t + nka)] \tag{I-5-3}$$

where k is a running index, n is the number in the lattice, A is amplitude, a is the lattice constant, and ω is an (angular) velocity.

Note from Fig. I-4b that these are trigonometric solutions. Also, if k has the value

$$k = (\pi/a)j \tag{I-5-4}$$

where j is an integer, then for $n = i$ and $n = i + 1$, Eq. (I-5-3) takes the values

$$\begin{aligned} u_{i+1} &= A \exp(i\omega t) \exp[i(n+1)ka] \\ &= A \exp(i\omega t) \exp(inka) \exp[i(\pi/a)ja] \\ &= u_i \exp(i\pi j) = -u_i \end{aligned} \tag{I-5-5}$$

Hence for the value of k given by Eq. (I-5-4) the solution is a standing wave. For other values of k, the solution is a traveling wave. Inserting Eq.

(I-5-3) into Eq. (I-5-2) now gives

$$-MA^2 \exp(i\omega t + kan) = AB \exp(i\omega t)\{\exp[ika(n+1)] + \exp[ika(n-1)] - 2\exp(ikan)\} \quad \text{(I-5-6)}$$

Dividing by $A \exp(i\omega t) \exp(ikan)$ then gives

$$-M\omega^2 = B[\exp(ika) + \exp(-ika) - 2] = B[2\sin(ka/2)]^2 \quad \text{(I-5-7)}$$

i.e.,

$$\omega = 2(B/M)^{1/2} \sin(ka/2) \quad \text{(I-5-8)}$$

Note that for values of ka in the interval

$$-\pi < ka < \pi \quad \text{(I-5-9)}$$

the value of $\sin ka/2$ will increase from -1 to $+1$ and the values will be unique. Values of k in the interval

$$-\pi/a < k < +\pi/a \quad \text{(I-5-10)}$$

are referred to as the first Brillouin zone (Brillouin, 1931). If the value of k is larger π/a by an amount x (where $x < \pi/a$), then

$$\omega - 2(B/M)^{1/2} \sin(ka/2) = 2(B/M)^{1/2} \sin(\pi/2 + x)$$
$$= 2(B/M)^{1/2} \sin(\pi/2 - x)$$

and this argument is in the interval given by Eq. (I-5-10); i.e., the properties of the propagated wave have already been discussed. In this case one may consider part of the wave as propagated, part as reflected.

For order-of-magnitude calculations one might ask what would be the value of B, for instance, at 60 μm, recalling that the region where lattice vibrations have an effect on the IR spectrum is between 20 and 60 μm. For $\lambda = 60 \times 10^{-4}$ cm

$$\omega = 2\pi\nu = 2\pi(3 \times 10^{10})/(60 \times 10^{-4}) = 3.14 \times 10^{13} \quad \text{sec}^{-1} \quad \text{(I-5-11)}$$

If the molecular weight of the compound is 200, then

$$M = 200/(6 \times 10^{23}) = 3.3 \times 10^{-22} \quad \text{(I-5-12)}$$

so that

$$3.14 \times 10^{13} = 2[B/(3.3 \times 10^{-22})]^{1/2} \quad \text{(I-5-13)}$$

or

$$B = 5.2 \times 10^4 \quad \text{dyn/cm} \quad \text{(I-5-14)}$$

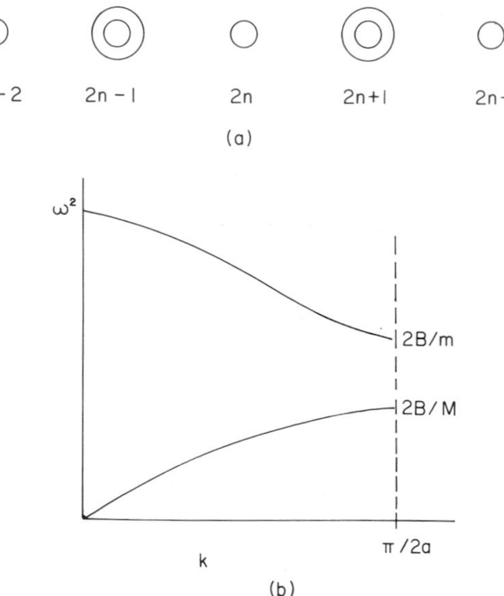

Fig. I-5 (a) Positions of ions in a 1 : 1 electrolyte. (b) ω^2 as a function of k showing the forbidden band (first Brillouin zone).

The elastic stiffness coefficient is given by

$$c = B/a \quad \text{dyn/cm}^2 \tag{I-5-15}$$

since there are $1/a$ molecules per centimeter. Inserting $a = 5 \times 10^{-8}$ as a value for the lattice constant, one obtains

$$c = (5.2 \times 10^4)/(5 \times 10^{-8}) \sim 10^{12} \quad \text{dyn/cm}^2 \tag{I-5-16}$$

For CO_2 (Walmsley, 1968) $c = 10^{11}$ dyn/cm^2, and for NaCl (Kittel, 1956, p. 60) $c = 5 \times 10^{11}$ dyn/cm^2, so that there is fair order-of-magnitude agreement.

For ionic crystals the situation will be as shown in Fig. I-5a. Here positive ions of mass m will occupy the even-numbered sites ($2n-2, 2n, 2n+2$), and negative ions of mass M will occupy the odd-numbered sites. The governing differential equations will in this case be

$$m\ddot{u}_{2n} = B(u_{2n+1} + u_{2n-1} - 2u_{2n}) \tag{I-5-17}$$

$$M\ddot{u}_{2n+1} = (u_{2n+2} + u_{2n} - 2u_{2n-1})B \tag{I-5-18}$$

Solutions to this are sought in the form

$$u_{2n} = E\exp[i(\omega t + 2nka)] \tag{I-5-19}$$

$$u_{2n+1} = F\exp\{i[\omega t + (2n+1)ka]\} \tag{I-5-20}$$

When Eqs. (I-5-19) and (I-5-20) are inserted in Eqs. (I-5-17) and (I-5-18), the resulting equations are[†]

$$-m\omega^2 E = BF[\exp(ika) + \exp(-ika)] - 2BE \qquad \text{(I-5-21)}$$

$$-M\omega^2 F = BE[\exp(ika) + \exp(-ika)] - 2BF \qquad \text{(I-5-22)}$$

For these equations to have nontrivial solutions the determinant must equal zero, which allows one to calculate ω^2 as a function of k:

$$\omega^2 = B[(1/m) + (1/M)] \pm B\{[(1/m) + (1/M)]^2 - 4\sin^2(ka)/Mm\}^{1/2} \qquad \text{(I-5-23)}$$

For small values of k this reduces to

$$\omega^2 = 2B[(1/m) + (1/M)] \qquad \text{(I-5-24)}$$

$$\omega^2 = 2Bk^2a^2/(M+m) \qquad \text{(I-5-25)}$$

In the first of these one gets, upon insertion into Eq. (I-5-21) that

$$E/F = -M/m \qquad \text{(I-5-26)}$$

i.e., the ions vibrate in directions opposite to one another. By a method similar to the one described previously, one can show that the limiting value is 52 μm, which is in good agreement with 60 μm found for NaCl for a stiffness coefficient of 5×10^{11} dyn/cm^2 and a lattice constant of 3×10^{-8} cm.

The two values of ω^2 given by inserting $\sin ka = 1$ in Eq. (I-5-23) are

$$\omega^2 = B[(1/m) + (1/M)] \pm B[(1/m) - (1/M)] = 2B/m \quad \text{or} \quad 2B/M \qquad \text{(I-5-27)}$$

The curves in the entire range of k in the first Brillouin zone are shown in Fig. I-5b. Note that there is a band of ω^2 values that are not permitted, i.e.,

$$2B/m < \omega^2 < 2B/m \qquad \text{(I-5-28)}$$

Using the value of 5×10^{11} dyn/cm^2 for c, and $a = 3 \times 10^{-8}$ cm, and inserting the masses for Na^+ and Cl^-, one finds this forbidden band to be

$$2.25 \times 10^{13} < \omega^2 < 2.74 \times 10^{13} \qquad \text{or} \qquad 69\ \mu\text{m} < \lambda < 84\ \mu\text{m} \qquad \text{(I-5-29)}$$

Using the value of c just quoted for NaCl in Eq. (I-5-24) gives $\lambda = 52$ μm, so that NaCl should have a fairly sharp absorption maximum between 52 and 69 μm, which is close to experimental fact (Kittel, 1956, p. 60). The

[†] For absorption of light, a term of Ce, where C is the electric field strength, should be added. This does not affect the limiting value.

above phenomena are the reason that the alkali halides are fairly transparent up to about 60 μm and can therefore be used as matrices for IR pellet work.

I-6 POLYMORPHISM

It was seen earlier that ionic crystals crystallize in crystal systems that are governed by the ratio of the ionic radii. In the case of molecular crystals, for instance, where organic molecules are spaced in the lattice positions, one obviously does not have the same degree of symmetry as one has in a fairly spherical ion, and therefore the stacking arrangement of these molecules cannot be expected to be as uniquely defined as for ionic substances. Indeed, if one had an organic molecule of the shape shown in Fig. I-6a, these L-shaped molecules could, for instance, be visualized in

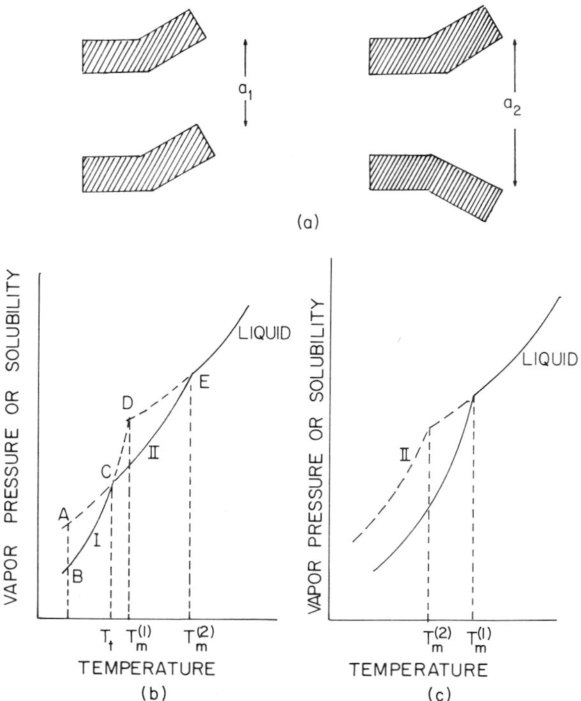

Fig. I-6 (a) Example of molecular arrangements of the same molecular entity in two different polymorphs. (b) Pressure or solubility curve of a substance crystallizing in two polymorphs with a transition point (enantiotropic polymorphs). (c) Pressure or solubility curve of a substance without transition point (monotropic polymorphs).

two logical arrangements, denoted configurations 1 and 2. Note that the lattice constants a_1 and a_2 are different, and that therefore the lattices will be different. Indeed most organic molecules can crystallize in several different ways, and these different lattice configurations of the same molecule are called *polymorphs*.

It is obvious that these different configurations would be associated with different energies, and that therefore one configuration would be more stable than other configurations. Since the "unstable" polymorphs are energetically unfavored, one would expect them to change spontaneously into the stable configuration. The energy of activation of this transformation, however, is quite high, and, especially when kept dry, the "unstable" modifications can "keep" indefinitely and are therefore denoted *metastable* polymorphs.

Although many different polymorphs may exist, the discussion to follow will compare a stable and a metastable polymorph. The metastable polymorph may have a higher energy than the stable polymorph at any temperature below the melting temperature, as shown in II of Fig. I-6c. In this case one talks about a *monotropic* polymorph. On the other hand, if a stable modification is heated, it may be that at one temperature T_t it will transform into another form, which is the stable form at $T > T_t$. In this case (Fig. I-6b) one talks about *enantiotropic forms* and the temperature T_t is denoted the transition temperature. Note from Fig. I-6c that in the case of a monotropic polymorph the melting temperature $T_m^{(2)}$ is always lower than that of the stable polymorph ($T_m^{(1)}$). In the case of enantiotropic forms, the form which is stable below T_t has the lower melting point. Note that owing to the high activation energy of transition one can easily "pass through" the transition point C, i.e., heat the form which is stable at lower temperature through the transition point, and obtain the melting point D. Slow heating can achieve the transition at C and give the ("real") melting point at E.

There is no accepted nomenclature for polymorphs, e.g., a subscript 1, I, or A does not imply a stable form or 2, II, or B a metastable form in the next higher energy state, and so on. The reason for this lack of nomenclature is that frequently when a drug, or compound, is developed and crystallized out from particular solvents, a particular crystal modification may be obtained. There is no way of knowing whether this is really the form with the lowest energy, and it frequently occurs (e.g., in the development of diazepam) that during the initial development of a drug a metastable monotropic form is produced. When, at a later (e.g., scale-up) stage a more stable form is produced (e.g., *the* stable polymorph), it may at times be impossible to produce the metastable form, since seeds of the stable form (e.g., airborne) are present.

Polymorphism is usually checked by one of six methods:

(1) x-ray diffraction,
(2) polarized microscopy,
(3) IR spectra,
(4) melting points,
(5) solubility measurements, and
(6) differential scanning calorimetry (or differential thermal analysis).

In x-ray diffraction patterns one can obviously obtain lattice spacings (and in single-crystal diffraction, atomic and group coordinates), and hence distinguish between polymorphs. In the polarized microscope one can distinguish crystal system and axis angles, and therefore one has a tool for distinguishing between crystals which differ in these respects. IR spectra in the high-wavelength region have already been described; since the spectrum is a function of the lattice, different lattice constants will give different spectra, and hence this provides a means of distinguishing between different polymorphs. Melting points can, as mentioned, be used to establish differences in morphology, and in such techniques one frequently melts, cools, and remelts. This may (at times) give different melting points, the last melting point usually belonging to the (high-temperature) stable modification.

Solubility is also used as a criterion for polymorphism. As shown in Fig. I-7a, one can determine the solubility at different temperatures, and plot according to

$$\ln S = \ln S_0 - (\Delta H/R)(1/T) \qquad \text{(I-6-1)}$$

where S is solubility in milligrams (or grams) of solute per gram (or kilogram) of solvent, R the gas constant, T absolute temperature, S_0 a constant, and ΔH a heat of solution term.

The solubility of a compound is usually determined by placing an excess of solid in contact with the solvent at constant temperature (e.g., in a constant temperature bath) and providing agitation (e.g., by use of rotating bottles). The concentration of the supernatant is then determined at, e.g., 12, 24, 48, and 72 hr, or until a steady level is reached. For compounds that are stable in solution this is an adequate method, except that in the case of metastable polymorphs one may first approach the metastable solubility S_2; but nucleation of the stable polymorph is a distinct possibility, and in spite of metastability in the solid state, conversion in "wet" state is much more easily accomplished. Such precipitation often occurs prior to the achievement of the solubility of the metastable polymorph, so that the concentration versus time profile will have the

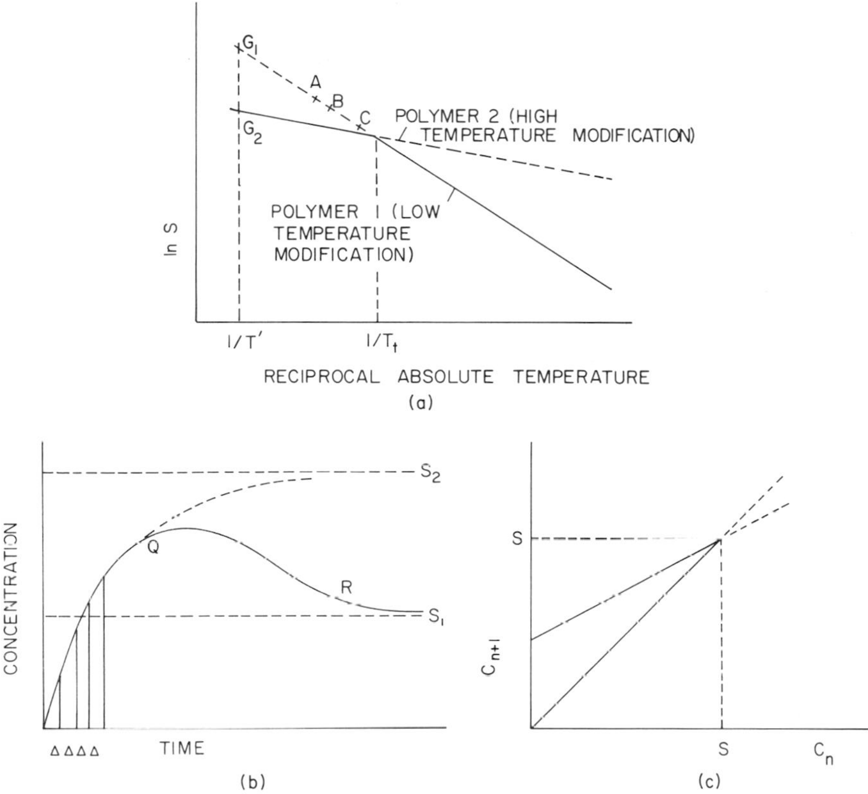

Fig. I-7 (a) Logarithmic representation of the solubility of two enantiotropic polymorphs, as a function of inverse absolute temperature. (b) Determination of the solubility of the metastable polymorph by the method of (Nogami *et al.*, 1966). (c) Amount dissolved as a function of time during dissolution of a metastable polymorph.

shape shown in Fig. I-7b. The point Q indicates where crystallization of the stable form starts, and the curve QR is frequently an exponential decay curve.

A method for obtaining S_2 from the first part of the curve has been developed by Nogami *et al.* (1966). Concentrations in the supernatant are obtained at equal intervals Δ, and the concentrations denoted C_1, $C_2, \ldots, C_n, \ldots, C_j$. When one plots C_{n+1} versus C_n a straight line ensues (Fig. I-7c) for the following reason: as will be demonstrated later, when the surface area does not change during a dissolution process (as with excess of solid where only small surface area changes result),

$$C_\theta = S[1 - \exp(-k\theta)] \tag{I-6-2}$$

At times θ and $\theta + \Delta$ this gives

$$C_\theta = S[1 - \exp(-k\theta)] \tag{I-6-3}$$

and

$$C_{\theta+\Delta} = S\{1 - \exp[-k(\theta + \Delta)]\} = S[1 - \exp(-k\Delta)] + \exp(-k\Delta)\, C_\theta \tag{I-6-4}$$

The solubility can be obtained graphically from the point where the C_{n+1} versus C_n line intersects $C_{n+1} = C$, and algebraically from the slope and intercept. The latter procedure in addition gives the value for k. Solubilities obtained in this fashion, of course, do not have the precision of equilibrium solubilities (as, e.g., S_1 in Fig. I-7b).

Again making reference to Fig. I-7a, the Gibbs energy of solid polymorph 1 at absolute temperature T' is the same as the Gibbs energy of a saturated solution at that temperature and is denoted G_1:

$$G_1 = G_1^0 + RT \ln S_1 \tag{I-6-5}$$

Similarly the Gibbs energy of solid polymorph 2 is the same as that of a saturated solution and is G_2, given by

$$G_2 = G_2^0 + RT \ln S_2 \tag{I-6-6}$$

If one denotes by a the activity of the dissolved species, then since $G = G^0 + RT \ln a$ for the solutions through the entire (ideal) concentration range, it follows that the reference values G^0 are the same in Eqs. (I-6-5) and (I-6-6), so that

$$G_2 - G_1 = \Delta G = RT \ln(S_2/S_1) \tag{I-6-7}$$

Since the solids are in equilibrium with the solutions of concentrations S_1 and S_2, their Gibbs energies are equal to those of the solutions, and hence Eq. (I-6-7) is an expression for the difference in energy between the two

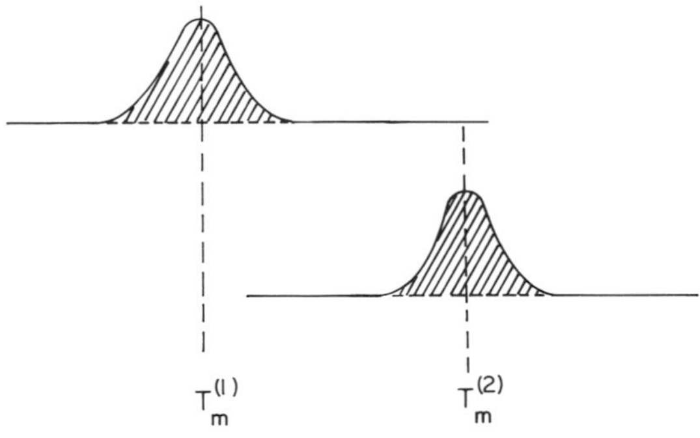

Fig. I-8 Melting point endotherms for two polymorphs.

polymorphs. It should be noted parenthetically that Eq. (I-6-1) is a direct consequence of Eq. (I-6-5), since

$$[\partial(\Delta G/T)/\partial T]_p = -\Delta H/T^2 = R(\partial \ln S/\partial T)_p \qquad \text{(I-6-8)}$$

which integrates directly to Eq. (I-6-1).

The last point regarding polymorphism is that it is frequently checked by DSC. This method consists of exposing a sample to a programed temperature (i.e., a preset temperature time curve). The amount of heat added to the sample (as compared to a control) is plotted. Typical traces for the two polymorphs in Fig. I-6b, c are shown in Fig. I-8. The heat of fusion is the area under the curve and it follows that

$$\Delta H_2 = \Delta H_1 + c_p(T_m^{(2)} - T_m^{(1)}) \qquad \text{(I-6-9)}$$

where c_p is the heat capacity of the melt.

I-7 CRYSTAL DEFECTS

Crystals are never perfect; in fact, even single crystals have a mosaic structure and consist of crystallites with boundaries. This means that when one performs x-ray diffraction studies on single crystals not only the thermal motion of the lattice points but also the quoted imperfections will cause broadening of bands which should otherwise be quite sharp. Aside from this, however, the individual crystallites will not be perfect, and the most common defects are screw dislocations and vacancies.

In a screw dislocation (Fig. I-9) the crystal grows in a spiral; therefore there is no need for any other nucleation than the initial nucleation for the crystal to *grow*. In other types of crystal growth the growth *rate* is a function of surface nucleation (Mullin, 1961) and when surfaces grow over one another there is a possibility of creating vacancies in the crystal.

Vacancy types are shown in Fig. I-10a–d. In ionic crystals they are usually Schottky or Frenkel defects (Fig. I-10d). The former one can visualize as the result of moving a pair of ions from the interior of the crystal to the surface or of introducing a higher valent ion. In this last case, the introduction of one Ca^{++} ion in a KCl lattice will create one vacancy,

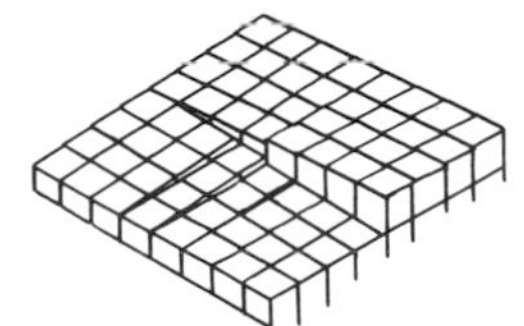

Fig. I-9 Screw dislocation in a crystal.

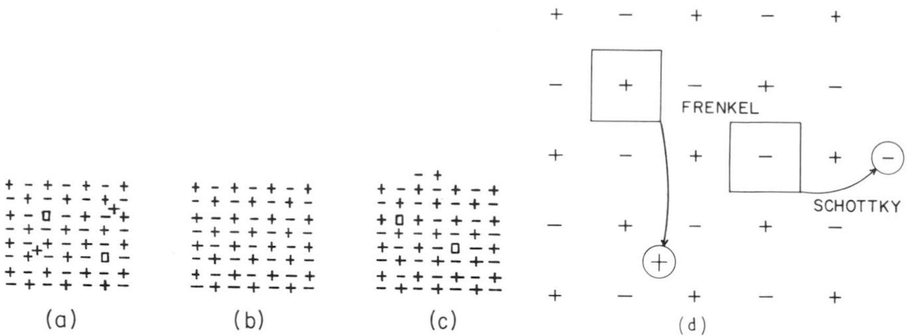

Fig. I-10 Formation of defects: (a) lattice vacancies and interstitial ions, (b) normal lattice, (c) lattice vacancies with migration of ions to the surface, and (d) schematic drawing of the formation of Frenkel and Schottky defects.

as shown in Fig. I-11a. If the change in density is plotted as a function of the concentration of Ca^{++} ions, one can calculate what it should be (Fig. I-11b), as opposed to what it actually is. The theoretical change in density is obtained by considering that one Ca^{++} ion requires one lattice site and one vacancy and weighs 40 atomic units, i.e., 20 units per site, whereas two K^+ ions occupy two sites, i.e., 39 units per site. One can therefore from the density calculate the concentration of vacancies.

There is a certain amount of "equilibrium" vacancies in a crystal; the following treatment parallels that of Kittel (1956, p. 478).

Reference will be made to Fig. I-12. It may be seen that in general if one forms n vacancies by moving molecular units to the surface, then the number of ways of making up the crystal is given by

$$Q = \binom{N+n}{n} = (N+n)!/N!\,n! \tag{I-7-1}$$

The entropy of a system is k, the Boltzmann constant, times the number of ways in which the system can be made up, i.e.,

$$S = k \ln Q = k \ln\left[(N+n)!/N!\,n!\right] \tag{I-7-2a}$$

This expression can be simplified by use of Stirling's formula:

$$\ln x! \sim x \ln x - x \tag{I-7-2b}$$

so that

$$S/k = (N+n)\ln(N+n) - N \ln N - n \ln n \tag{I-7-3}$$

With the crystal considered as a constant-volume system, equilibrium is achieved when the Helmholtz energy is at minimum, i.e., when $dF/dn = 0$. If E denotes the energy of transfer of one molecular unit to the surface, one may write the Helmholtz energy as

$$F = E - TS = nE - kT \ln Q \tag{I-7-4}$$

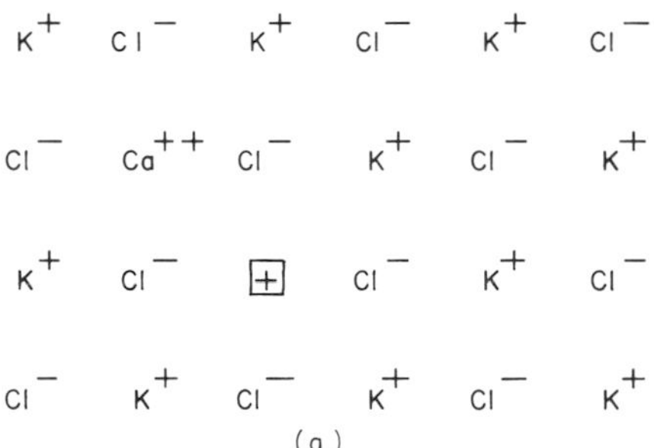

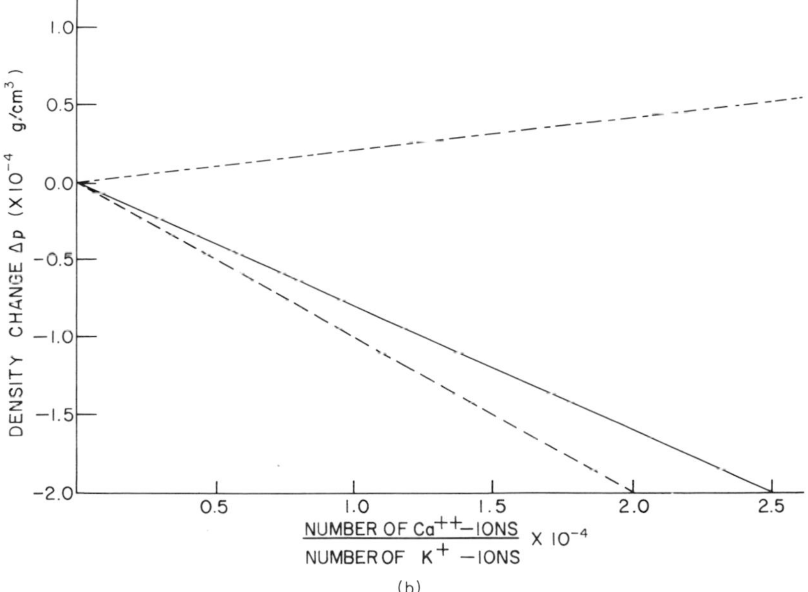

Fig. I-11 (a) Formation of vacancies in potassium chloride by inclusion of calcium ions; (b) density of crystal if no vacancies are allowed for (upper broken line), if vacancies are allowed for (lower broken line), and experimental curve (solid line). [After Pick and Weber (1950).]

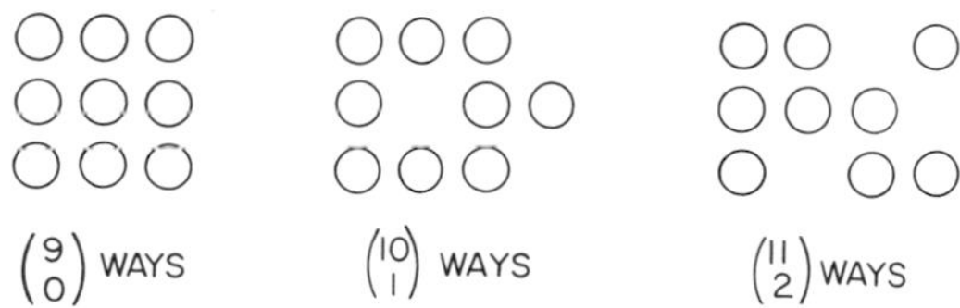

Fig. I-12 The number of ways in which one molecule can be moved from the interior to the surface and the number of ways in which two molecules can be moved from the interior to the surface.

so that

$$\{\partial F/\partial n\}_{\mathrm{v}} = E - kT(\partial \ln Q/\partial n)_{\mathrm{v}}$$
$$= E - kT\ln[(N+n)/n] \sim E - kT\ln(N/n) = 0 \quad \text{(I-7-5)}$$

or

$$n = N\exp(-E/kT) \quad \text{(I-7-6)}$$

This applies to "equilibrium" at the stated conditions, but also to Schottky and Frenkel defects. In these latter cases one would have to consider E to be the energy for a pair of ions being moved to the surface or a pair of ions formed in the lattice.

For instance, one can by volumetric density determination vis-à-vis lattice unit volume times Avogadro's number calculate the number of vacancies in sodium chloride. This is $5 \times 10^5/\mathrm{cm}^3$ and 1 cm^3 is 4×10^{22} ions, and one can therefore write

$$5 \times 10^5 = 2 \times 10^{22}\exp(-E/kT) \quad \text{(I-7-7)}$$

from which

$$E = 46.4\ \text{kcal/"ion pair"} \quad \text{(I-7-8)}$$

One interesting consequence of vacancies in alkali halides is that if one exposes an alkali halide M_1X crystal to the vapor of an alkali metal M_2, then one obtains a colored crystal. The color depends on M_1X, not on M_2. This type of phenomenon is called a color center or an F center (F from *Farbe*).

The explanation may partly be sought in considering the system akin to a particle in a box. The M_2 atom penetrates the lattice and positions itself in a vacancy but then dissociates into M_2^+ and an electron, which diffuses to another vacancy. Recall [see, e.g., Pitzer (1953, pp. 31–33)] that the energy levels are given by

$$E = \alpha^2 h^2/8\pi^2 m \quad \text{(I-7-9)}$$

If one assumes an infinite "wall height" (potential barrier at the end of the box), then

$$\alpha = n\pi/a \quad \text{(I-7-10)}$$

holds, so that

$$E = n^2 h^2/8ma^2 \quad \text{(I-7-11)}$$

The difference between the energies associated with $n = 1$ and $n = 2$ is therefore

$$E_2 - E_1 = 3h^2/8ma^2 = h\nu \quad \text{(I-7-12)}$$

Inserting the values $h = 6.63 \times 10^{-27}$ erg sec, $m = 9.1 \times 10^{-28}$ g, and $a = 3 \times 10^{-8}$ cm then gives

$$\nu = 0.03 \times 10^{17} \quad \text{sec}^{-1} \tag{I-7-13}$$

or

$$\lambda = 3 \times 10^{10}/0.03 \times 10^{17} = 10^{-5} \text{ cm} = 100 \text{ nm} \tag{I-7-14}$$

This is in fair agreement with the actual value of 350 nm, considering the assumptions made above (one-dimensional box, infinite wall height, and rest mass of the electron).

I-8 CRYSTALLIZATION

Crystallization is usually the last step prior to the final grinding of a produced chemical. In this case it is often termed recrystallization, and its purpose is either to purify the compound (by leaving impurities in the mother liquor) or to achieve better or different crystal shapes, sizes, or polymorphs. In recrystallization one may use one of the following methods:

(1) "salting out,"
(2) cooling a melt, or
(3) cooling a saturated solution of the substance in a solvent.

In case (1), one makes a solution of the compound (drug) DH in a solvent A and then either

(a) adds a different solvent B of different polarity (B is usually miscible with A, but DH is less soluble in A + B than in A);

(b) or if DH is protolytic, makes a solution of DH_2^+ (e.g., an amine salt if DH is an amine) or D^- (e.g., a salt of a carboxylic acid if that applies) in a polar solvent, and then adjusts the pH to the pK of the substance or beyond, so that DH (which is usually less soluble in the polar solvent) will precipitate out,

(c) or "salts" out by saturating the solution either with common or uncommon ions; one here relies on the ionic product or the ionic strength–solubility relationship, respectively.

In case (2) one melts the compound to be purified and cools. If the melt is not pure DH, the crystals appearing above the eutectic temperature (provided the impurities are present in amounts less than the eutectic composition) are pure DH, provided the impurities do not form solid solutions with the compound. This method is not generally used.

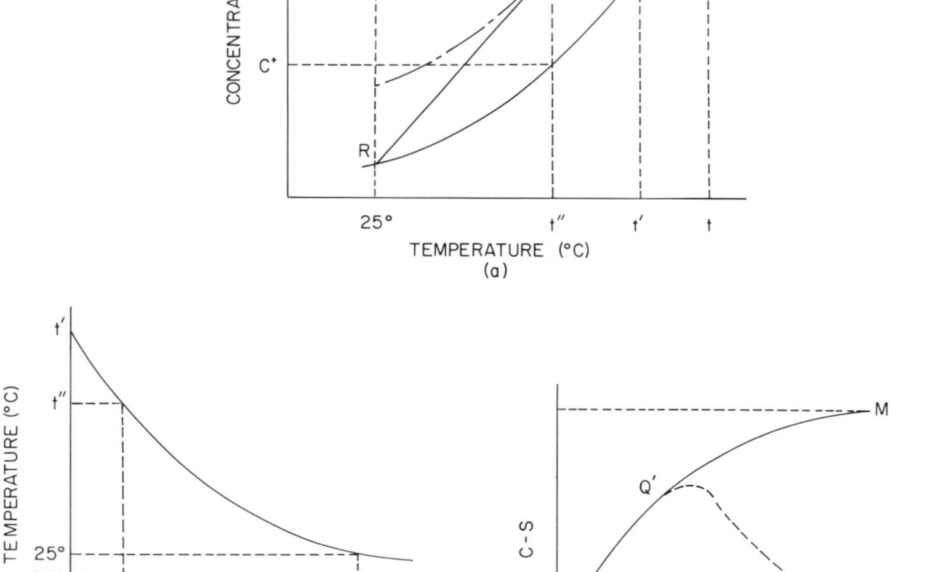

Fig. I-13 (a) Saturation curve QR of a substance as a function of temperature. If a solution at concentration C' is reduced from temperature t to t', it is saturated. Precipitation will occur at point Q′ at temperature t''. C_p is the size of the metastable zone. (b) Temperature–time curve when a solution of concentration C' is cooled from t' through t'' to 20°C. Precipitation occurs at time 0 and is carried out to time θ'. (c) Supersaturation as a function of time. Precipitation starts at time 0 (Q′) and ends at time τ (R).

Method (3) is commonly used and will be subject to some discussion and comments below, where we refer to Figs. I-13–I-19.

Suppose that DH is soluble in solvent A as shown in Fig. I-13a. If a mass (in milligrams) C' of drug is added to 1 g of solvent at 25°C and heated, all the material will be going into solution at temperature t' (°C). The solubility curve has the equation

$$S = S_0 \exp(-\Delta H/RT) \tag{I-8-1}$$

where ΔH is a heat of solution term, R is the gas constant, T is the absolute temperature (°K), S_0 is a constant, and S is the solubility in milligrams of solid per gram of solvent.

If the system (containing C' milligrams of solid per gram of solvent) is heated, the solid will go into solution at temperature t' (point Q). It is assumed that it is further "overheated" to a temperature of t and left there for time θ^*. The solution is then let cool by exposing it to an ambient (cooling) temperature of 20°, and it will be saturated at a temperature of t' (point Q). It will, however, supercool, and not until a temperature of t'' is reached will crystals start appearing. If this is done at other concentrations and the same heating history, i.e., the same value of θ^* and the same cooling rate are imposed, then the temperature of supercooling, denoted P, will be the same, that is,

$$P = t' - t'' \qquad \text{(I-8-2)}$$

and the zone between the bold and dotted line in Fig. I-13a is referred to as the *metastable zone*.

After the point Q′ has been reached, the temperature will decrease, e.g., by a curve such as that shown in Fig. I-13b. With the time θ measured beyond the point t' where crystallization starts, the equation for the temperature t versus time θ curve could be e.g.,

$$\ln(t - 20) = -K\theta + \ln(t' - 20) \qquad \text{(I-8-3)}$$

where K is a heat transfer coefficient and the ambient temperature is 20°C. For precipitation to occur the system must be supersaturated. The following two terms are used to describe the extent of this:

$$q = C/S, \quad \text{supersaturation ratio} \qquad \text{(I-8-4a)}$$

$$\Delta = C - S, \quad \text{degree of supersaturation} \qquad \text{(I-8-4b)}$$

As mentioned, the length of time a solution is kept at t (Fig. I-13a), the overheating, affects the width of the metastable zone. Fig. I-14 exemplifies a situation (Myl, 1960) in which a solution of KH_2PO_4 has been overheated by 50°C for varying lengths of time, and shows the effect of this on the width P of the metastable zone.

The degree of supersaturation of the system in Fig. I-13a is shown in Fig. I-13c. The solid curve QM shows Δ as a function of time with the (artificial) assumption that no precipitation occurs. However, as shown precipitation occurs at point Q′ ($\theta = 0$) and the supersaturation curve could be, e.g., as shown by the dashed curve Q′R, i.e., with the supersaturation decreasing with time. Fig. I-13c is labeled with nomenclature consistent with Fig. I-13a. For instance, the initial supersaturation, which is usually denoted $C_0 - S$, equals $C' - C^*$. Curve Q′R in Fig. I-13c is frequently an exponential decay.

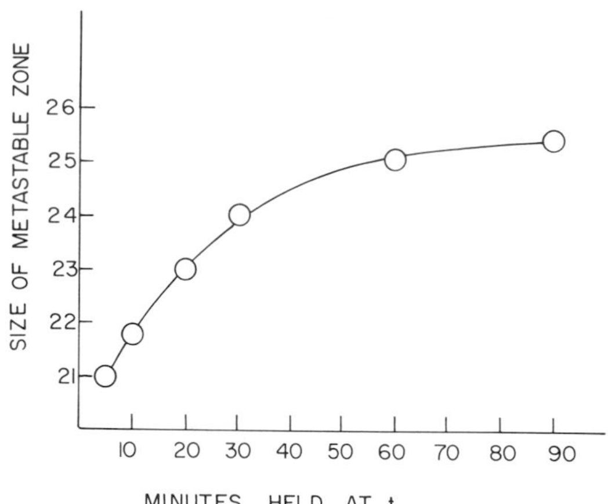

Fig. I-14 Size of the metastable zone for KH_2PO_4 as a function of overheating by 50° [After Myl (1960). Reproduced with permission of the copyright owner.]

The first step in crystallization is the formation of a nucleus. In the metastable zone molecular aggregates may form, but they are "destroyed" by thermal motion prior to achieving a size that will allow growth of a crystal. If the critical number of molecules in an aggregate necessary for growth is denoted n, one calls this n-aggregate a nucleus. Nuclei then appear from time $\theta = 0$ on and they are formed at a rate of

$$dN/d\theta = J \quad \text{nuclei/sec} \tag{I-8-5}$$

Many authors quote this quantity to be proportional to $C - S$, i.e.,

$$dN = J(C - S)\,d\theta \tag{I-8-6}$$

but it will be seen shortly that this is somewhat oversimplified.

Once a nucleus is formed, the crystal will grow at a growth rate $dm/d\theta$ which is larger with increasing (a) surface area A (cm^2) of the crystal and (b) degree of supersaturation, i.e.,

$$dm/d\theta = M(C - S)A \tag{I-8-7}$$

where M is the growth rate constant (Mullin, 1961). These two equations will apply simultaneously during the cooling curve.

We shall first consider the energy changes in the crystallization process in order to arrive at a physical concept of what is meant by the word nucleus. The molecules in a supersaturated solution are in random motion as shown in Fig. I-15, but finite probabilities exist for spatial arrangements congruent with the final lattice arrangement. For simplicity, the following argument dealing with a one-dimensional crystal is presented at this point. With reference to Fig. I-15, the probability of a dimer (2-aggregate)

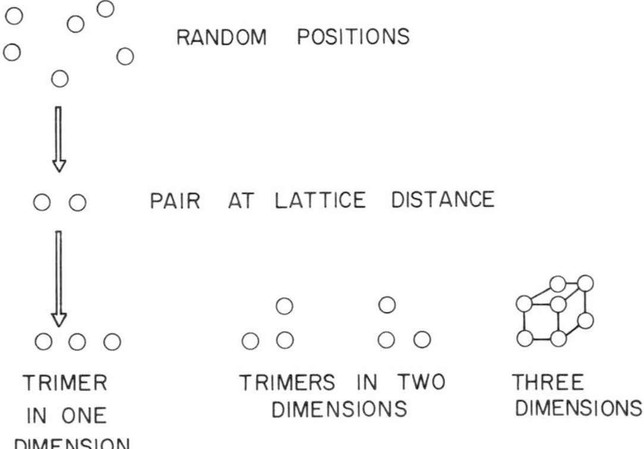

Fig. I-15 Steps from random movement to ordered nuclei in a one-dimensional crystal. Analogies in two and three dimensions are shown.

forming in a solution with a concentration (in mole fraction) of x is

$$P(2) = x^2/N^* \tag{I-8-8}$$

where N^* is the total number of all aggregate combinations possible. The probability of a trimer is

$$P(3) = 2x^3/N^* \tag{I-8-9}$$

because there are two positions to which the third molecule can "attach" itself. In a similar fashion, the probability of a j-aggregate is

$$P(x_j) = (j-1)x^j \Big/ \sum_{i=1}^{\infty} (i-1)x^i \sim jx^j \Big/ \sum_{i=1}^{\infty} ix^i \tag{I-8-10}$$

Note that the total N^* is given as a summation with infinity as upper limit. This is not strictly true but is used in lieu of enumerating the number of combinations. In a similar fashion the probability of an aggregate with more than n 1 molecules is given by

$$P(x \geqslant x_n) = \sum_{i=n}^{\infty} ix^i \Big/ \sum_{i=1}^{\infty} ix^i \tag{I-8-11}$$

The shapes of the curves represented by Eq. (I-8-11) are shown in Fig. I 16.

If we now assume that a cubical crystal with side r, i.e., of volume r^3 and surface $6r^2$, is formed, then the change in Gibbs energy associated with the transfer of one mole of substance from supersaturated solution to solid crystal is

$$\Delta G = G_s - G_v = 6r^2\sigma - r^3\mu \tag{I-8-12}$$

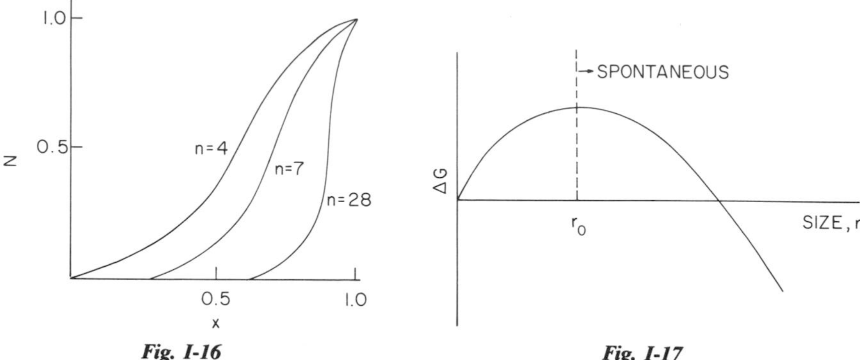

Fig. I-16 *Fig. I-17*

Fig. I-16 Curves of Eq. (I-8-11) for various values of *n*.

Fig. I-17 Free energy change as a function of the size (radius) of the nucleus.

where G_s, the energy associated with the production of the surface, is positive, since it requires the interfacial tension σ (erg/cm^2) to produce 1 cm^2 of surface, where G_v is negative, and where μ is the Gibbs energy per unit volume of crystal. The trace of Eq. (I-8-12) has the shape shown in Fig. I-17. Note that it first rises, but beyond the maximum an increase in r will be associated with a negative ΔG, so that the process is spontaneous. The value r_0 corresponds to the critical size, i.e., the size of the nucleus. It can be found by equating $d\,\Delta G/dr$ with zero:

$$d\,\Delta G/dr = 0 = 12r\sigma - 3r^2\mu = 0 \tag{I-8-13}$$

i.e.,

$$r_0 = 4\sigma/\mu \tag{I-8-14}$$

which inserted in Eq. (I-8-12) gives

$$\Delta G = 32\sigma^3/\mu^2 \tag{I-8-15}$$

The curve implies that a larger crystal (i.e., with side r_2) is more stable than a smaller crystal (with side r_1), as long as these lengths are larger than the critical size (r_0). Hence the former should have a lower solubility (i.e., $S_2 < S_1$). This so-called Ostwald–Freundlich equation (Ostwald, 1900; Freundlich, 1922) has been utilized by many authors (Higuchi, 1958; Smolen and Kildsig, 1971; Krause and Kildsig, 1972; Jeannin *et al.*, 1975) but has also been refuted (Bikerman, 1970). The argument leading to Eq. (I-8-24) is as follows. Consider a large and a small crystal such as shown in Fig. I-18 (here considered to be cubic for simplicity). By dissolving a mass dm_1 from the small crystal the area decreases by

$$dA_1 = 6r_1^2 - 6(r_1 - dr_1)^2 = 12r_1\,dr_1 \tag{I-8-16}$$

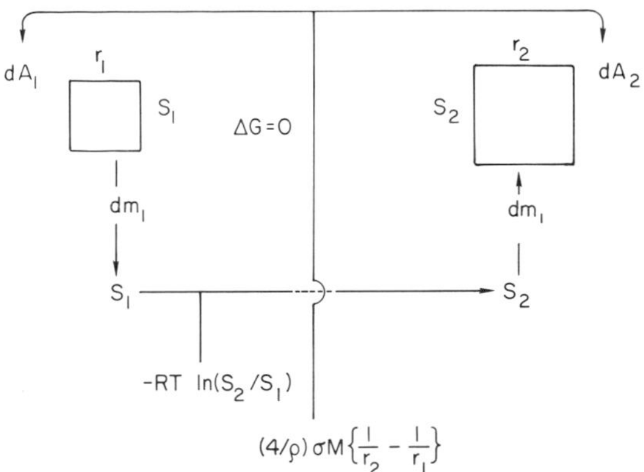

Fig. I-18 Energetic scheme for derivation for Eq. (I-8-24).

The mass of a cube is $m = \rho V = \rho r^3$, so that

$$dm_1 = 3\rho r^2\,dr = (\rho r_1/4)12r_1\,dr_1 = (\rho r_1/4)\,dA_1 \qquad \text{(I-8-17)}$$

Hence

$$dA_1 = (4/\rho r_1)\,dm_1 \qquad \text{(I-8-18)}$$

The mass dm_1 precipitates on the large cube and causes a change in area given by

$$dA_2 = (4/\rho r_2)\,dm_1 \qquad \text{(I-8-19)}$$

Hence the change in surface by the transfer of mass dm_1 is

$$dA_2 - dA_1 = (4/\rho)(1/r_2 - 1/r_1)\,dm_1 \qquad \text{(I-8-20)}$$

so that the work done by the surroundings on the system for the transfer of 1 mole (M rather than dm_1) is

$$U = (4M\sigma/\rho)(1/r_2 - 1/r_1) \qquad \text{(I-8-21)}$$

which is a negative number. The work done by the surroundings on the system to transfer one mole from S_1 to S_2 is

$$G = -RT\ln(S_2/S_1) \qquad \text{(I-8-22)}$$

which is positive (if $S_2 < S_1$). For microreversibility the overall energy change must be zero, so that

$$(4M\sigma/\rho)(1/r_2 - 1/r_1) = RT\ln(S_2/S_1) \qquad \text{(I-8-23)}$$

or

$$\ln(S_2/S_1) = (4M\sigma/RT\rho)[(1/r_2) - (1/r_1)] \qquad \text{(I-8-24)}$$

The term surface energy has been used extensively in the above treatment. A few words regarding its nature may be in order. Brunauer

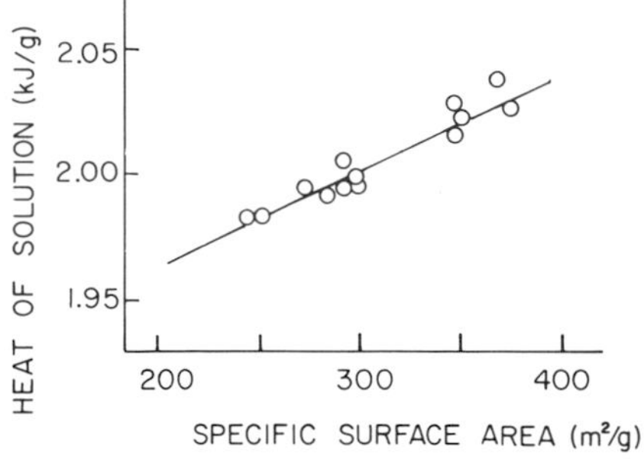

Fig. I-19 Variation of the enthalpy of solution of $Ca_3Si_2O_7$, $2H_2O$ with variation of specific surface area (1 g samples). [After Brunauer (1961). Reproduced with permission of the copyright owner.]

(1961) notes that surface energy is in essence the difference between the total energy of a molecule of a given substance in the surface layer and the total energy of the same molecule located in the interior of the crystal. If a solid substance is subjected to heat of solution measurements in various degrees of subdivision, then the heat of solution may be plotted versus the surface area of the substance, and a straight line ensues the slope of which, when appropriate units are used, is the surface energy. In essence what is obtained is the surface enthalpy, but the pv term is small, so that this is approximately equal to the surface energy. Figure I-19 shows data reported by Brunauer.

The chemical energy involved in the transfer of 1 mole from a solution of concentration C to one of concentration S is given by

$$\Delta G = -nRT \ln(C/S) \qquad \text{(I-8-25)}$$

where $n = 1$. Since a solution of concentration S is in equilibrium with the crystal, Eq. (I-8-25) represents the energetics of transfer from solution to solid state. Generally n is of order $\frac{1}{3}$–1 (Nyvit, 1971), and it shall be assumed that it is $\frac{1}{2}$ or 1. The ratio C/S is called the supersaturation ratio and the symbol q used for it below. If Eq. (I-8-25) represents the energy of activation for nucleation, then

$$J = dN/d\theta = J_0 \exp(-E/RT) = J_0 \exp(n \ln q) = J_0 q^n \qquad \text{(I-8-26)}$$

where use of Eq. (I-8-25) has been made in going from the second to the third step. If $C \gg S$, this may be written as

$$J = \alpha(C - S)^n \qquad \text{(I-8-27)}$$

The question of particle size distribution is one of concern to the pharmaceuticist and the powder technologist (Higuchi and Hiestand, 1963). For this reason a short discussion (Carstensen and Sheridan, 1979) of the generation of a particle size distribution from first principles of crystal growth will be presented here.

It will be assumed that the growth rate is of the form

$$dm/d\theta = MA\Delta^2 \tag{I-8-28}$$

where m is the mass, M is the growth rate constant, and A is the surface area of the crystal (Misra and White, 1971). We shall also assume as mentioned earlier that during crystallization the degree of supersaturation is a function of time given by

$$\Delta = (C_0 - S)\exp(-k\theta) \tag{I-8-29}$$

It should be noted that k is *not* a growth rate constant.

The crystallization starts at $\theta = 0$ and ends at $\theta = \tau$ (Fig. I-13c). We shall now attempt to find the number of crystals N_a formed at time $\theta = \Theta$ and the size to which they grow between times Θ and τ. We shall address ourselves to the last point first.

The mass m of a single cubic particle of side r is

$$m = r^3\rho \tag{I-8-30}$$

so that

$$dm/d\theta - 3r^2\rho(dr/d\theta) \tag{I-8-31}$$

The surface area A is

$$A = 6r^2 \tag{I-8-32}$$

Combining Eqs. (I-8-28)–(I-8-32) then gives

$$dm/d\theta = 3r^2\rho(dr/d\theta) = M6r^2(C_0 - S)^2\exp(-2k\theta) \tag{I-8-33}$$

or

$$dr/d\theta = (2k/a)\exp(-2k\theta) \tag{I-8-34}$$

where

$$a = 2\rho k/2M(C_0 - S)^2 \tag{I-8-35}$$

If the particle has a birth time of Θ and is let grow until time τ, then

$$r_\Theta = (2k/a)\int_\Theta^\tau \exp(-2k\theta)\,d\theta = (1/a)[\exp(-2k\Theta) - \exp(-2k\tau)] \tag{I-8-36}$$

The largest crystal is the one whose birth time is zero (Fig. I-14); its side length is

$$r_0 = (1/a)[1 - \exp(-2k\tau)] \tag{I-8-37}$$

Hence

$$r_0 - r_\Theta = (1/a)[1 - \exp(-2k\Theta)] \tag{I-8-38}$$

or

$$\exp(-2k\Theta) = 1 - a(r_0 - r_\Theta) = ar_\Theta + b \tag{I-8-39}$$

where

$$b = 1 - ar_0 \tag{I-8-40}$$

A result which will be used later is that Eq. (I-8-34) can now be written as

$$d\theta = (a/2k)(ar_\Theta + b)^{-1}\,dr \tag{I-8-41}$$

The number of crystals formed at time θ is dN and is the nucleation rate at that particular time:

$$J = dN/d\theta = \alpha[C(\theta) - S]^{1/2} = \alpha(C_0 - S)^{1/2}\exp(-k\theta/2) \tag{I-8-42}$$

where n in Eq. (I-8-26) is taken as 0.5. The number N nuclei formed in a time τ (in seconds) of crystallization experience is given by

$$\begin{aligned} N &= \alpha\int_0^\tau (C_0 - S)^{1/2}\exp(-k\theta/2)\,d\theta \\ &= (2\alpha/k)(C_0 - S)^{1/2}[1 - \exp(-k\tau/2)] \end{aligned} \tag{I-8-43}$$

The fraction having sizes between r_Θ and $r_\Theta + dr_\Theta$ [Eq. (I-8-36)] obtained by combining Eqs. (I-8-42) and (I-8-43):

$$dN/N = k\exp(-k\Theta/2)\,d\theta/2[1 - \exp(-k\tau/2)] \tag{I-8-44}$$

Inserting Eqs. (I-8-39) *and* (I-8-34) into this gives

$$\begin{aligned} dN/N &= k(ar_\Theta + b)^{0.25}(a/2k)(ar_\Theta + b)^{-1}/2[1 - \exp(-k\tau/2)]\,dr_\Theta \\ &= Q(ar_\Theta + b)^{-3/4}\,dr_\Theta \end{aligned} \tag{I-8-45}$$

If $b \ll a$, then

$$dN/N = Q(ar_\Theta)^{-3/4}\,dr_\Theta \tag{I-8-46}$$

Integrating from 0 to r_0 then gives

$$1 = 4Q(ar_0)^{1/4} \qquad \text{or} \qquad Q = 0.053a^{-1/4} \tag{I-8-47}$$

so that the fraction of particles smaller than r is given by

$$\int_0^r (0.053/a)r_\Theta^{-3/4}\,dr_\Theta = (0.21/a)r^{1/4} \tag{I-8-48}$$

This type of distribution resembles the log-normal distribution for 80% of the particle distribution.

REFERENCES

Bikerman, J. J. (1970). "Physical Surfaces," p. 216. Academic Press, New York.

Born, M., and Huang, K. (1954). "Dynamic Theory of Crystal Lattices," p. 15. Oxford Univ. Press, London.

Brillouin, L. (1931). "Die Quantenstatistik," p. 72. Springer-Verlag, Berlin and New York.

Brunauer, S. (1961). *In* "Solid Surfaces and the Gas–Solid Interface,"(L. E. Copeland, ed.), p. 15, Adv. Chem. Ser. No. 33. American Chemical Society, Washington, D. C.

Budnikov, P. P., and Ginstling, A. M. (1968). "Principles of Solid State Chemistry," pp. 8–11. Maclaren and Sons, Ltd., London.

Carstensen, J. T. (1973). "Theory of Pharmaceutical System," Vol. II, pp. 95–99. Academic Press, New York.

Carstensen, J. T. and Sheridan, J. (1979). *Acta Pharm. Technol.* **25**, 127 (1979).

Davies, M. (1959). *J. Polym. Soc.* **40**, 247.

Davies, M. (1971). *J. Chem. Educ.* **48**, 591.

Davies, M., and Jones, A. H. (1959). *Trans. Faraday Soc.* **55**, 1329.

Davies, M., and Kybett, B. (1963). *Nature (London)* **200**, 776.

Davies, M., and Kybett, B. (1965). *Trans. Faraday Soc.* **61**, 1608, 1893.

Davies, M., and Malpass, V. E. (1961). *J. Chem. Soc.* (6), 1048.

Davies, M., and Thomas, G. H. (1960). *Trans. Faraday Soc.* **56**, 185.

Davies, M., Jones, A. H., and Thomas, G. H. (1959). *Trans. Faraday Soc.* **55**, 1100.

Ferro, D. R., and Hermans, J., Jr. (1972). *Biopolymers* **11**, 105.

Freundlich, H. (1922). "Colloid and Capillary Chemistry," p. 155. Dutton, New York.

Higuchi, T. (1958). *J. Am. Pharm. Assoc. Sci. Ed.* **47**, 657.

Higuchi, W., and Hiestand, E. (1963). *J. Pharm. Sci.*, **52**, 67.

Jeannin, C., Pothisiri, P., and Carstensen, J. T. (1975). *Ann. Pharm. Fr.* **33**, 433.

Kittel, C. (1956). "Introduction to Solid State Physics," pp. 9, 60, 478. Wiley, New York.

Krause, P. D., and Kildsig, D. O. (1972). *J. Pharm. Sci.*, **61**, 281.

Misra, C., and White, E. T. (1971). *Adv. Chem. Eng.* **64**, 53.

Mullin, J. W. (1961). "Crystallization," pp. 3–29. Butterworths, London.

Myl, J. (1960). *Sb. Pr. V. U. An. Chem. Usti*, No. 4, 21.

Nogami, H., Nagai, T., and Suzuki, A. (1966). *Chem. Pharm. Bull.* **14**, 329.

Nyvit, J. (1971). "Industrial Crystallisation from Solutions," pp. 34–51. Chemical Rubber Publ. Co., Cleveland, Ohio.

Ostwald, W. (1900). *Z. Phys. Chem.* **34**, 503.

Pellegrini, A., Ferro, D. R., and Zerbi, G. (1973). *Mol. Phys.* **20**, 577.

Pick, H., and Weber, H. (1950). *Z. Phys.* **128**, 409–413.

Pitzer, K. S. (1953). "Quantum Chemistry," 5th ed., pp. 31–33, 44, 81–82, 215, 415. Prentice-Hall, Englewood Cliffs, New Jersey.

Shuttleworth, S. (1949). *Proc. Phys. Soc. London* **A62**, 167.

Slater, J. C. (1924). *Phys. Rev.* **23**, 488.

Smolen, V., and Kildsig, D. O. (1971). *J. Pharm. Sci.* **60**, 130.

Walmsley, S. H. (1968). *J. Chem. Phys.* **48**, 1438.

CHAPTER

II

Particulate Solids

If a solid exists as a group of individual particles, then the properties of the population as a whole, *the particulate system*, often depend on individual particle parameters, such as, e.g., their size. In many applications it is convenient to consider the particles of which a particulate solid consists to be spherical. This will be done on many occasions in this book, but at the onset it should be pointed out that only very few *pure* substances are present as spherical particles. Substances such as starch are fairly spherical, but most crystalline inorganic or organic salts have irregular shape. On the other hand, processed particles, such as granules, are often pseudospherical.

II-1 SHAPE AND DIMENSIONS

In general it is possible to determine both volumes and surface areas of geometrical solids which are of fairly regular shape. But when the particles are not regular, the problem becomes more difficult; in fact, it is frequently difficult to define the volumes and surface areas in geometric terms. In cases like this one often makes reference to a particular dimension, for instance, the length, the thickness, or the width.

II-1-1 Particle Dimensions

The words "the diameter of a particle" will frequently be utilized either to describe a given dimension (as stated above) or in accord with some particular definition. It is very common to define the diameter d'_v of a particle as the diameter of the sphere which has the same volume V as the particle:

$$d'_v = \left[(6/\pi)V\right]^{1/3} \tag{II-1-1}$$

In one gram of monodisperse powder, i.e., a powder where all the particles are of the same "size," d'_v would be given by

$$d'_v = (6/\pi\rho_p N)^{1/3} \tag{II-1-2}$$

where N is the number of particles per gram and ρ_p is the particle density.

In analogy with Eq. (II-1-1) it is possible to define a particle diameter as the diameter d'_s of the sphere which has the same surface area A as the particle:

$$d'_s = (A/\pi)^{1/2} \tag{II-1-3}$$

II-1-2 Individual Particle Shapes

There is, obviously, a need to define particle shape, and many attempts have been made in this area. Some of these tie in with polydisperse powders and will be discussed later, but the following can be defined from single particles (or monodisperse powders):

The *rugosity* or sphericity Γ (Wadell, 1934) is defined as the ratio between the actual surface area A of the particle and the surface area A' of a sphere with the same volume:

$$\Gamma = A/A' \tag{II-1-4}$$

Note that this ties in directly with the diameter d'_s (Eq. II-1-3).

Another shape factor, introduced by Dallavalle (1948), is the *circularity* ϕ, which relates the cross-sectional perimeter C of the particle (such as, e.g., obtained by microscopy) with the circumference C' of a circle with the same area,

$$\phi = C/C' \tag{II-1-5}$$

Example II-1-1 The shape shown in Fig. II-1 is the perpendicular cross section of a cylinder 10 μm high. Calculate d'_v, ϕ, and Γ.

Answer The cross-sectional area is estimated (graphically) to be $70 - 27 = 43$ μm^2. A circle with an area of 43 μm^2 has a diameter of $[(4/\pi)43]^{1/2} = 7.4$ μm. The circumference of this is $7.4\pi = 23.2$ μm. The perimeter of the cross section in Fig. II-1 is estimated at 40 μm, so that

$$\phi = 40/23.2 = 1.7$$

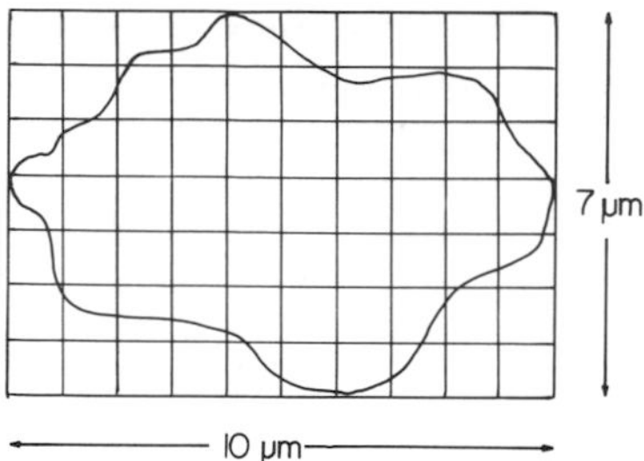

Fig. II-1 Particle dimensions of an irregular particle.

The surface area of the *particle* is $40 \cdot 10 + 2 \cdot 43 = 486\ \mu m^2 = A$. The volume of the particle is $V = 10 \cdot 43 = 430\ \mu m^2$, so that a sphere of this volume has a diameter of

$$d_v' = [430/(\pi/6)]^{1/3} = 9.4\ \mu m$$

This sphere has a surface area of

$$A' = \pi 9.4^2 = 277\ \mu m^2$$

so that

$$\Gamma = A/A' = 486/277 = 1.75.$$

II-2 MEAN PARTICLE DIAMETERS

Real particulate systems usually consist of particles of varying diameters. There are occasions where products are truly monodisperse (e.g., various pollens), but in most cases the diameters will not all be identical, in which case the powder is said to be *polydisperse*. There will be a distribution of these diameters, and the most logical first thought in defining a *mean diameter* is to count the number n_i of particles with diameter d_i and to proceed in this fashion until all the particles in the powder sample have been classified. The *arithmetic mean diameter* is then given by

$$d_n = \sum_{i=1}^{N} n_i d_i \Big/ \sum_{i=1}^{N} n_i \qquad \text{(arithmetic mean diameter)} \qquad \text{(II-2-1)}$$

The limits ($i = 1$ and N) denote the smallest and largest diameter, and for convenience of presentation these limits will not be included in subsequent formulas (unless necessary). Hence n_i is the number of particles with diameter d_i, and in the following notation simply n and d will be used.

Counting methods of various types exist; microscopy, for instance, gives d_n directly, as demonstrated in the following example. It should be noted that, traditionally, at least ten fields and at least (a total of) 200 particles are counted. Even so, the sampling is a problem, and the question of how representative it is must be considered.

Example II-2-1 A microscopic scan gives the following results for a powder sample: 10 particles below 2 μm, 34 particles between 2 and 5 μm, 48 particles between 5 and 10 μm, 64 particles between 10 and 25 μm, 30 particles between 25 and 50 μm, 9 particles between 50 and 75 μm, 5 particles between 75 and 100 μm, and none above 100 μm. What is the arithmetic mean diameter?

Answer Table II-1 is constructed as shown. The column marked $n_i d_i$ is derived by assuming d_i to be equal to the class midpoint (shown in the column headed d_i). Hence $d_n = 3938/200 = 19.69 = 20$ μm.

The term mean diameter is, however, by no means unambiguous. For example, definitions of mean diameter via Eqs. (II-1-1) and (II-1-3) are possible, in which case one obtains (Dallavalle, 1948)

$$d_v = \left(\sum nd^3/\sum n\right)^{1/3} \quad \text{(mean volume diameter)} \qquad \text{(II-2-2)}$$

$$d_s = \left(\sum nd^2/\sum n\right)^{1/2} \quad \text{(mean surface diameter)} \qquad \text{(II-2-3)}$$

where d_v is the diameter *directly* obtained by Coulter counter determinations.

TABLE II-1

First, Second, Third and Fourth Diameter Moments of a Particle Size Distribution

Class interval (μm)	Interval midpoint d_i (μm)	Number of particles[a] n_i	$n_i d_i$[a]	$n_i d_i^2$	$n_i d_i^3$[a]	$n_i d_i^4$	Cumulative Percent by Number n_i	Cumulative Percent by Weight $n_i d_i^3$
< 2	1	10 (5%)	10	10	10	10	5	
2–5	3.5	34 (17%)	119	417	1458 (0.02%)	5102	22	0.02
5–10	7.5	48 (24%)	360	2700	20250 (0.28%)	151875	46	0.30
10–25	12.5	64 (32%)	800	10,000	125,000 (1.72%)	1,562,500	78	2.02
25–50	37.5	30 (15%)	1125	42,188	1,582,031 (21.74%)	59,326,172	93	23.76
50–75	62.5	9 (4.5%)	562.5	35,156	2,197,265 (30.20%)	137,329,102	97.5	53.96
75–100	87.5	5 (2.5%)	438	38,281	3,349,609 (46.04%)	293,090,820	100	100
Totals		200 (100%)	3938	128,752	7,275,623	491,465,581		

[a] Number in parenthesis is noncumulative percent.

There are yet other mean diameters, those emerging naturally from various measurements. If the surface A of a powder and the true solid volume V are measured, then

$$V/A = (\pi/6)\sum nd^3/\pi\sum nd^2 \tag{II-2-4}$$

For a single particle $6V/A = d$ and hence $6V/A$ is a type of diameter. The expression in Eq. (II-2-4) is denoted the surface volume mean diameter and is given by

$$d_{\mathrm{sv}} = \sum nd^3/\sum nd^2 \qquad \text{(surface volume mean diameter)} \tag{II-2-5}$$

This is the diameter directly obtained by permeametry measurements.

It follows, too, that if one obtains the weights w_i of particle fractions of a particle diameter d_i, then a natural diameter is the weight average:

$$d_{\mathrm{vm}} = \sum w_i d_i/\sum w_i \qquad \text{(weight (or volume) mean diameter)} \tag{II-2-6}$$

but since $w_i = n_i(\pi/6)\rho d_i^3$, it follows that

$$d_{\mathrm{vm}} = \sum nd^4/\sum nd^3 \tag{II-2-7}$$

This is the mean diameter obtained directly from sieve analysis and from the Andreasen apparatus (sedimentation analysis).

Example II-2-2 Find d_{v}, d_{s}, d_{sv}, and d_{vm} from the data in Example II-2-1.

Answer The sums of the first, second, third, and fourth moments are listed in Table II-1. The diameters are as follows:

$$d_{\mathrm{vm}} = 491{,}465{,}581/7{,}275{,}623 = 67.5 \quad \mu\mathrm{m}$$

$$d_{\mathrm{sv}} = 7{,}275{,}623/128{,}752 = 56.5 \quad \mu\mathrm{m}$$

$$d_{\mathrm{v}} = (7{,}275{,}623/200)^{1/3} = 33.1 \quad \mu\mathrm{m}$$

$$d_{\mathrm{s}} = (128{,}752/200)^{1/2} = 25.4 \quad \mu\mathrm{m}$$

Detailed experimental techniques are beyond the scope of this book, but the following general statements are in order.

In *microscopy* one measures a particle dimension and obtains a number frequency curve; hence the directly obtained particle diameter is d_n (II-2-1).

In *electronic counting* (Coulter counter) one measures the number of times a sample contains particles with volume larger than a threshold volume, and hence it is a cumulative *distribution* of volume versus number. One therefore directly obtains the mean volume diameter (II-2-2), which most often is converted arithmetically to volume mean diameter (II-2-6).

In *sedimentation methods* (Andreasen pipette) one obtains directly the percent by weight of particles with diameters larger than a certain size, i.e., the data are cumulative by weight, and hence a *distribution* is obtained and the mean diameter is the weight mean diameter (II-2-6).

As mentioned, *sieving* gives the weight mean diameter [Eq. (II-2-6)] directly as a *frequency* plot.

Finally, *permeametry* gives the surface volume mean diameter indirectly (since it measures surface area per cubic centimeter of solid) and does not give distributional data.

II-3 PARTICLE SIZE DISTRIBUTIONS

The previous sections have dealt with polydispersicity in a general sense, accounting for the fact that a certain fraction of particles will have one diameter and a certain fraction another, but the actual distribution of these diameters is important in solid pharmaceutics.

II-3-1 The Log-Normal Distribution

Particle size distributions in naturally occurring processes are frequently log-normally distributed. This, for instance, is the case with isothermally crystallized materials, with spray-dried granulations, and with milled particles. It will be assumed in this book that powders are log-normally distributed by number, as in the following example.

Example II-3-1 Plot the data in Example II-2-1 appropriately. Find the *mean* (d_g) and the *standard deviation* ($\ln \sigma^2$).

Answer Figure II-2 shows the percentages by weight and the percentages by number plotted as a function of the logarithm of the particle diameter. It is obvious from the figure that the number distribution is a straight line and hence that the particle sizes are distributed log-normally by number. It follows directly from the figure (for 50%) that the mean diameter is 11 μm. The mean diameter is denoted d_g when it refers to a number distribution and d_g^w when it refers to a weight distribution.

To find the standard deviation, recall that 84% of all measurements are below the mean plus one standard deviation, and as indicated in the figure, 84% corresponds to a diameter of 30. However, the standard deviation for log-normal distribution is a logarithmic number and hence is denoted $\ln \sigma$, which is found as follows:

$$\ln \sigma = \ln d(84\%) - \ln d_g = \ln 30 - \ln 11 = 3.401 - 2.398 = 1.003$$

The standard deviation of course could have been found as the difference between $\ln d_g$ and $\ln d(16)$ as well.

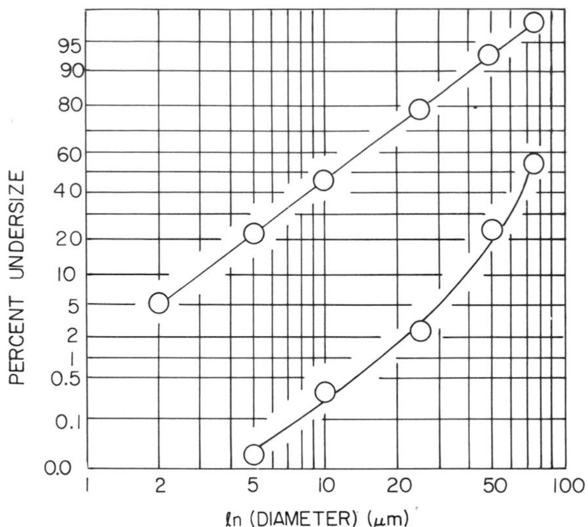

Fig. II-2 Data from Example II-2-1 plotted as percentage by weight and percentage by number.

Note from Fig. II-2 that the number distribution is log-normal, i.e., is a straight line, whereas the weight distribution is not a straight line. It is common to obtain the mean and the standard deviation from the appropriate distribution, in this instance from the number distribution, and then to calculate the means for the other types of distributions from this particular set of numbers. This is known as the Hatch–Choate theorem (Hatch and Choate, 1929). Prior to describing this, however, we should point out that the mean diameter obtained from a log-normal distribution d_g is a geometric mean, because

$$\ln d_g = \left(\sum n_i \ln d_i\right)/\left(\sum n_i\right) = \ln\left(\prod_{i=1} d^{n_i}\right)^{1/(\sum n_i)} \tag{II-3-1}$$

A normal distribution obeys the following equation:

$$f(d) = \left(1/\sigma\sqrt{2\pi}\right)\exp\left[-(d-\bar{d})^2/2\sigma^2\right] \tag{II-3-2}$$

where $f(d)$ is the fractional frequency of the diameter d (with mean $\bar{d}$). A log-normal distribution (Dallavalle, 1948) follows

$$f(\ln d) = \frac{1}{\ln\sigma\sqrt{2\pi}}\exp\left[-\frac{(\ln d - \ln d_g)^2}{2\ln^2\sigma}\right] \tag{II-3-2a}$$

where d_g is the geometric mean diameter and $f(\ln d)$ is the number of particles having diameters d between $\ln d$ and $\ln d + \partial \ln d$.* For a distribution which is log-normal as far as numbers are concerned, we may therefore write

$$d_n = \frac{\sum nd}{\sum n} = \frac{\sum n}{\sum n(\ln\sigma)\sqrt{2\pi}} \int_0^{\infty} d \exp\left[-\frac{(\ln d - \ln d_g)^2}{\sqrt{2}\,\ln\sigma} \right] \partial \ln d \quad \text{(II-3-3)}$$

The following substitution is now introduced:

$$u = (\ln d - \ln d_g)/(\sqrt{2}\,\ln\sigma) \quad \text{(II-3-4)}$$

Note that $0 < d < \infty$ implies $-\infty < u < +\infty$. It may be seen from Eq. (II-3-4) that

$$du = \left[1/(\sqrt{2}\,\ln\sigma)\right] \partial \ln d \quad \text{(II-3-5)}$$

from which it follows that

$$\ln(d/d_g) = (\sqrt{2}\,\ln\sigma)u \quad \text{(II-3-6)}$$

or

$$d = d_g \exp(\sqrt{2}\,\ln\sigma)u \quad \text{(II-3-7)}$$

Hence Eq. (II-3-3) becomes

$$d_n = \frac{\sqrt{2}\,\ln\sigma}{(\ln\sigma)\sqrt{2\pi}} \int_{-\infty}^{+\infty} d_g \exp\left[(\sqrt{2}\,\ln\sigma)u - u^2\right] du \quad \text{(II-3-8)}$$

Now, since

$$(\sqrt{2}\,\ln\sigma)u - u^2 = -\left[u - (\sqrt{2}/2)\ln\sigma\right]^2 + \tfrac{1}{2}\ln^2\sigma \quad \text{(II-3-9)}$$

the substitution

$$m = u - (\sqrt{2}/2)\ln\sigma \qquad (dm = du) \quad \text{(II-3-10)}$$

can be inserted in Eq. (II-3-8) to yield

$$\begin{aligned} d_n &= \left(d_g/\sqrt{\pi}\right)\int_{-\infty}^{+\infty} \exp(\tfrac{1}{2}\ln^2\sigma)\exp(-m^2)\,dm \\ &= \left(d_g/\sqrt{\pi}\right)\exp(\tfrac{1}{2}\ln^2\sigma) \times 2(\sqrt{\pi}/2) = \exp(\tfrac{1}{2}\ln^2\sigma + \ln d_g) \end{aligned} \quad \text{(II-3-11)}$$

Similar expressions can be obtained for the other types of diameters and these are tabulated in Table II-2.

* The symbol ∂ is used for differentials with diameters to avoid combinations such as *dd*.

TABLE II-2

Hatch–Choate relations

$\ln d_n = \ln d_g + 0.5 \ln^2 \sigma$	(II-3-11′)
$\ln d_s = \ln d_g + \ln^2 \sigma$	(II-3-12)
$\ln d_v = \ln d_g + 1.5 \ln^2 \sigma$	(II-3-13)
$\ln d_{sv} = \ln d_g + 2.5 \ln^2 \sigma$ hr[a]	(II-3-14)
$\ln d_g = \ln d_g^w + 2.5 \ln^2 \sigma$ hr[b]	(II-3-15)
$\ln d_n = \ln d_g^w - 2.5 \ln^2 \sigma_w$	(II-3-16)
$\ln d_s = \ln d_g^w - 2 \ln^2 \sigma_w$	(II-3-17)
$\ln d_v = \ln d_g^w - 1.5 \ln^2 \sigma_w$	(II-3-18)
$\ln d_{sv} = \ln d_g^w - 0.5 \ln^2 \sigma_w$	(II-3-19)

[a] Obtained as the difference between $\ln d_v^3$ and $\ln d_s^2$.

[b] The geometric mean diameter is d_g^w when the distribution is log normal by weight, and has a standard deviation of $\ln \sigma_w$. The geometric mean diameter is d_g when the distribution is log-normal by number and has a standard deviation of $\ln \sigma$

II-3-2 Specific Shape Factors

It has been mentioned that specific shape factors may be defined as

$$s = \alpha_s d_s^2 \tag{II-3-12}$$

and

$$v = \alpha_v d_v^3 \tag{II-3-13}$$

where the lower case s and v denote individual particle surface and volume, respectively. If a unit weight of powder contains N particles, then

$$1/\rho N = \alpha_v d_v^3 \tag{II-3-14}$$

Hence, the specific surface area is

$$A_w = \alpha_s d_s^2 / \rho \alpha_v d_v^3 \tag{II-3-15}$$

Introducing Eqs. (II-3-12) and (II-3-13) into Eqs. (II-3-14) and (II-3-15) gives

$$-\ln N = \ln d_g^3 + 4.5 \ln^2 \sigma + \ln(\alpha_v \rho) \tag{II-3-16}$$

so that by experimentally obtaining the parameters other than α_v, this latter can be calculated.

Eq. (II-3-15) may also be rearranged to read

$$\ln A_w = -\ln \rho\alpha_v + \ln \alpha_s + \ln d_s^2 - \ln d_v^3.$$

From the above α_v is known, A_w and ρ can be determined experimentally,* and $\ln(d_s^2)$ and $\ln(d_v^3)$ are obtainable from the Hatch–Choate relations in Table II-2, so that

$$\ln A_w = -\ln \rho\alpha_v + \ln \alpha_s - \ln d_g - 2.5 \ln^2 \sigma \qquad \text{(II-3-17)}$$

Hence α_s can be obtained from the remaining parameter values.

Example II-3-2 From the data in Fig. II-2 find d_n, d_s, d_v, and d_{sv}, using the Hatch–Choate equations.

Answer The data are log-normal in a number distribution, and hence Eqs. (II-3-11)–(II-3-14) are used. $\ln d_g = \ln(11) = 2.398$. $\ln^2 \sigma = 1.006$. From Eq. (II-3-11), $\ln d_n = 2.398 + (0.5 \times 1.006) = 2.901$, so $d_n = 18.2$ μm. From Eq. (II-3-12), $\ln d_s = 2.398 + 1.006 = 3.404$, so $d_s = 30.1$ μm. From Eq. (II-3-13), $\ln d_v = 2.398 + (1.5 \times 1.006) = 3.907$, so $d_v = 49.8$ μm. These numbers differ somewhat from those found in Example II-2-2, which were based on class interval midpoints and hence are not as accurate as those stated above (which are rigorously based on the distribution being number log-normal).

II-4 SURFACE AREAS

Surface areas are measured by gas adsorption, permeametry, sedimentation, sieving, and counting means (microscopy and Coulter counter). The most direct method for measuring the surface is by gas adsorption. Nitrogen is the usual adsorbed gas, and two types of isotherms are of pharmaceutical importance.

II-4-1 The Langmuir Isotherm

In the following it is assumed that a solid with a clean surface is exposed to a gas. It is assumed that there are N sites on the surface at which the gas molecules can adsorb up to a monolayer, that adsorption of a molecule on any site is associated with the same energy E, and that this is independent of the degree of coverage. *All* these assumptions taken together are unlikely to occur at once, but the Langmuir isotherm does apply to several systems. Once equilibrium is established at a particular gas pressure P, the molecules will adsorb and desorb. Equation (II-4-1) can be derived by a kinetic (Langmuir, 1916) as well as by a thermodynamic

*Since the porosity is a function of size, ρ (in particular for granulations) may be a function of d, and this can introduce an error into this calculation. Note that ρ is the particle density (and not the true density).

(Volmer, 1925) or a statistical-mechanical approach (Fowler, 1935). Only the first will be presented here.

In the kinetic approach the rates of adsorption and desorption of a gas on a solid surface are used to derive the adsorption isotherm. Let k_- denote the rate of desorption, k_+ the rate of adsorption, and ζ their ratio. Then the desorption rate is $k_-\theta$, where θ is the fraction coverage, and $k_+P[1-\theta]$ is the rate of adsorption. At equilibrium, these must be equal, so

$$k_-\theta = k_+P(1-\theta) \qquad \text{or} \qquad \theta = \zeta P/(1+\zeta P)$$

or

$$1/\theta = 1 + 1/\zeta P \tag{II-4-1}$$

It is conventional to linearize the relation by some variant of Eq. (II-4-1) (Fig. II-3). If two gases (1 and 2) are present, then at equilibrium the now two equations become

$$\theta_1 = \zeta_1 P_1(1-\theta_1-\theta_2), \qquad \theta_2 = \zeta_2 P_2(1-\theta_1-\theta_2)$$

These two simultaneous equations yield the solution

$$\theta_1 = \zeta_1 P_1/(1+\zeta_1 P_1 + \zeta_2 P_2)$$

which can be simplified if $\zeta_1 P_1$ and $\zeta_2 P_2$ are of different magnitude:

$$1/\theta_1 = 1 + 1/\zeta_1 P_1, \qquad \zeta_1 P_1 \gg \zeta_2 P_2$$

$$\theta_1 = 1/(1+\zeta_2 P_2), \qquad \zeta_2 P_2 \gg \zeta_1 P_1$$

When $\zeta_2 P_2$ is considerably larger than unity, the latter term simply becomes $\zeta_1 P_1/\zeta_2 P_2$. Note that ζ is a function of the heat of adsorption (ΔH_a): $\zeta = A\exp(\Delta H_a/RT)$.

In experiments with glass beads and mica Langmuir (1916) found linear plots, but the volume of one monolayer was only 8–36% of the area

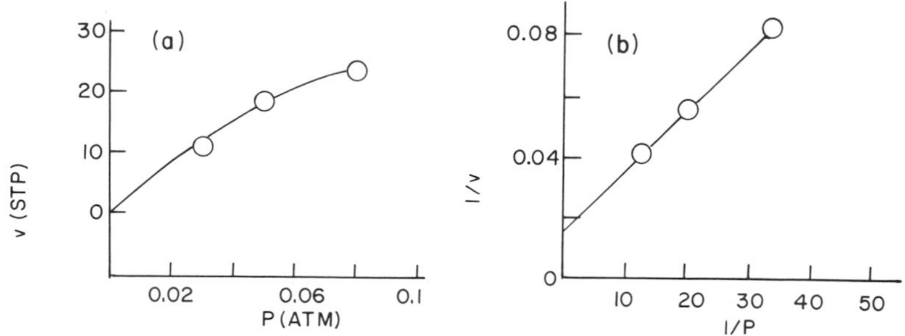

Fig. II-3 (a) Amount adsorbed as a function of equilibrium pressure (data from Table II-3). (b) Data plotted according to Eq. (II-4-1).

which was possible from geometrical considerations (and this, in turn, would be less than the actual area due to surface roughness). This is due to the invalidity of assuming that all the sites have identical energies. Many surfaces have as much as 20% of the total number of sites of higher adsorption affinity (larger ΔH) than the remainder. These may be cracks, Frenkel steps, or simply edges, and they are all completely covered, even at very low pressures.

Note in Eq. (II-4-1) that $\theta = N/N_\infty$, where N_∞ is the number of molecules in a monolayer. In general the amount or quantity measured is the volume of nitrogen adsorbed, v. This is converted to v at STP, and Eq. (II-4-1) becomes

$$v_\infty / v = (\zeta / P) + 1 \qquad \text{(II-4-1a)}$$

since v is proportional to N. One can then calculate v_∞, the reciprocal of the y intercept, from a graph of $1/v$ versus $1/P$. Of course, ζ can be calculated from the slope.

Example II-4-1 A 3 g sample gives the results listed in Table II-3. Calculate the surface area, assuming N_2 to be 16 Å^2 per molecule.

Answer A plot of the data will show that $1/v$ is linear in $1/P$, as is also obvious from inspection of the table; the equation for the line is

$$1/v = 0.002(1/P) + 0.017. \qquad \text{(II-4-2)}$$

Hence $v_\infty = 1/(0.017) = (59\ \text{cm}^3)/(3\ \text{g}) = (15.8 \times 10^{20}\ \text{molecules})/(3\ \text{g}) = (253 \times 10^4\ \text{cm}^2)/(3\ \text{g}) = 84.3\ \text{m}^3/\text{g}$. The data are plotted in Fig. II-3.

A Langmuir isotherm will have the general shape shown in Fig. II-4a. Two isotherms are shown, one at a temperature of T_1 °K and one at a temperature of T_2. In this fashion it is possible to calculate the *isosteric differential heat of adsorption* q (Jacobs and Tompkins, 1955), given by

$$q = RT^2(\partial \ln p / \partial T)_\theta \qquad \text{(II-4-3)}$$

TABLE II-3

Langmuir Adsorption Data[a]

Equilibrium pressure P (atm)	v (STP) (cm^3)	$1/P$	$1/v$
0.08	24	12.5	0.042
0.05	18	20	0.056
0.03	12	33.3	0.083

[a]Correlation coefficient, 0.999; intercept, 0.017; slope, 0.002.

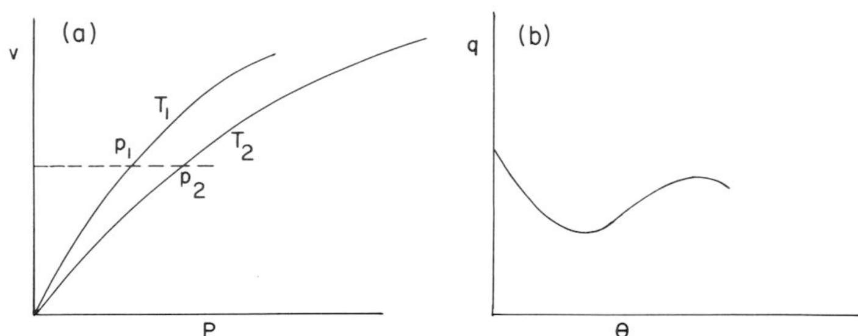

Fig. II-4 (a) Isotherms at two different temperatures, T_1 and T_2. (b) Data plotting according to Eq. (II-4-3a).

where p is the equilibrium pressure at a certain coverage θ (Fig. II-4a). By using two isotherms at T_1 and T_2, this of course becomes

$$-q = R(\ln p_2 - \ln p_1)/[(1/T_2) - (1/T_1)] \tag{II-4-3a}$$

If the degree of coverage were nonimportant (i.e., if this Langmuir assumption were correct) then q should be independent of θ, but plots such as shown in Fig. II-4b result (Jacobs and Tompkins, 1955). This is interpreted as follows: The initial decrease in q with θ is due to high-energy sites. Beyond this range, at low coverage, there is only neglible interaction between adsorbed molecules (their distance being great), leading to the (minimum) plateau. As $\theta \to 1$, the remainder of the sites are comparable to a uniform surface and the increase in q (i.e., the maximum) is due to attractive interaction between molecules which are getting closer and closer together. The final decrease in q is due to a *second adsorbed layer*.

II-4-2 The BET Isotherm

It may be seen from the above and from the fact that most isotherms have the shape shown in Fig. II-5 that most of the assumptions made for the Langmuir isotherm do not apply. A theory which accounts for multilayer adsorption is that of Brunauer *et al.* (1938).

In Fig. II-6 a situtation is depicted at a certain equilibrium pressure p. The area (in square meters) of the surface is A, and a part (s_0) of this is not covered, a part (s_1) is covered with a monolayer, and a part with a dilayer (s_2). For the general treatment below, a total of n layers is assumed. The adsorbed molecules occupy a volume v_0 per square meter of monolayer, then the volume of the first layer in Fig. II-6 is $v_0 s_1$, the volume of the second layer is $s_2 2v_0$, and the volume of the ith layer is $s_i i v_0$ (m^2).

The rate of adsorption on the uncovered surface (to form the first adsorbed layer) is proportional to the area (s_0) of the uncovered surface

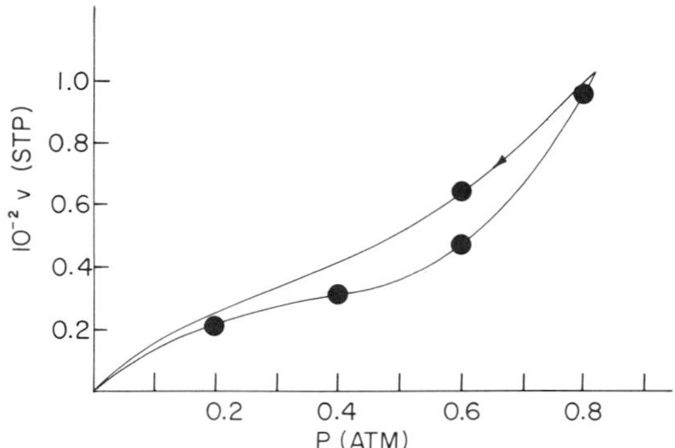

Fig. II-5 BET isotherm. (Data from Table II-4.)

and to the gas vapor pressure p, with a rate constant a_1 (the subscript 1 denoting the first sorbed layer); i.e., the rate is $a_1 s_0 p$. The rate of desorption is proportional to the area of the first adsorbed layer (s_1) with a rate constant of $k_1 = b_1 \exp(E/RT)$, where E is the energy of adsorption for the first layer. Hence at equilibrium, the following equation holds for the first layer:

$$a_1 s_0 p = b_1 s_1 \exp(-E/RT) \tag{II-4-4}$$

For the second layer to be formed, molecules adsorb onto the area s_1 (m^2) of the first, i.e., the rate is $a_2 s_1 p$, and the desorption is proportional to the area of the second layer (s_2) with a rate constant $b_2 \exp(-E_L/RT)$, i.e.,

$$s_2 b_2 \exp(-E_L/RT) = s_1 a_2 p \tag{II-4-5}$$

and similarly

$$s_3 b_3 \exp(-E_L/RT) = s_2 a_3 p \tag{II-4-6}$$

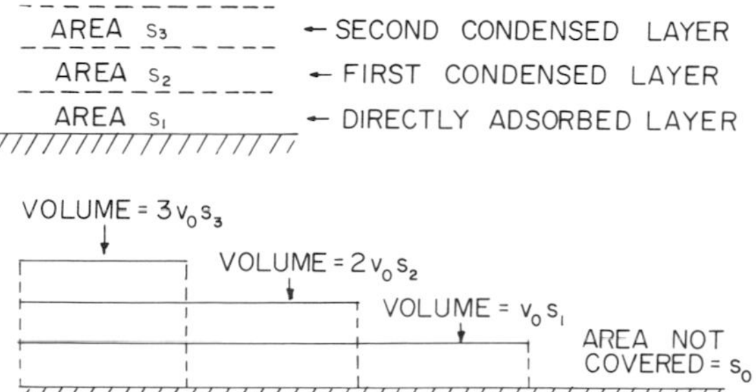

Fig. II-6 Schematic of mechanisms for BET isotherms.

It is assumed that E_L is the same for all layers (except the first) and is equal to the heat of vaporization of the adsorbing gas (the heat of liquefaction). The assumption is next made that in the general equation of types (II-4-5) and (II-4-6),

$$s_i b_i \exp(-E_L/RT) = s_{i-1} a_i p \tag{II-4-7}$$

the ratio a_i/b_i is constant for $i > 1$ (i.e., for layers other than the first absorbed layer). Hence Eq. (II-4-5) may be written

$$s_2 = (a/b) p \exp(E_L/RT) = x s_1 \tag{II-4-8}$$

where

$$x = p(a/b)\exp(E_L/RT) \tag{II-4-9}$$

where $a/b = a_i/b_i$ is the common value for the ratio. From Eq. (II-4-8) it is seen that, successively,

$$s_2 = x s_1 \tag{II-4-8}$$

$$s_3 = x s_2 = x^2 s_1 \tag{II-4-10}$$

and in general

$$s_i = s_1 x^{i-1} \tag{II-4-11}$$

Similarly, one may now introduce

$$s_1 = p s_0 (a_1/b_1)\exp(E/RT) = y s_0 \tag{II-4-12}$$

where

$$y = (a_1/b_1) p \exp(E/RT) \tag{II-4-13}$$

Denoting

$$y/x = c \tag{II-4-14}$$

we obtain

$$c = [(a_1/b_1)/(a/b)]\exp[(E - E_L)/RT] \tag{II-4-15}$$

and

$$s_1 = c s_0 x \tag{II-4-16}$$

The total area as shown in Fig. II-6 for a situation with two condensed layers is $s_0 + s_1 + s_2$, so that for n condensed layers it would be

$$\begin{aligned} A &= s_0 + s_1 + s_2 + \cdots + s_n = s_0 + s_0 c x + s_0 c x^2 + \cdots \\ &= s_0\left(1 + c\sum_{i=1}^{N} x^i\right) \end{aligned} \tag{II-4-17}$$

Again making reference to Fig. II-6, the volume of gas adsorbed in toto is

$$v = v_0(s_1 + 2s_2 + 3s_3 + \cdots + ns_n)$$
$$= v_0\left(cs_0x + 2cs_0x^2 + 3cs_0x^3 + \cdots\right) = v_0cs_0 \sum_{i=1}^{n} ix_i \qquad \text{(II-4-18)}$$

The quantity sought is v_m, the volume of the adsorbed monolayer, since this can be converted to number of molecules, each with an effective cross section, hence allowing evaluation of the area. It follows that

$$v_m = Av_0 = v_0s_0\left(1 + c\sum x^i\right) \qquad \text{(II-4-19)}$$

where Eq. (II-4-17) has been used in the last step of the equation. Combining Eqs. (II-4-18) and (II-4-19) then gives

$$\frac{v}{v_m} = \frac{v_0s_0c\sum ix^i}{v_0s_0\left(1 + c\sum x^i\right)} = \frac{c\sum ix^i}{1 + c\sum x^i} \qquad \text{(II-4-20)}$$

Note that $\sum x^i = x/(1-x)$ since it is a geometric progression. Also note that $\sum ix^i = x(\partial\sum x^i/\partial x) = x\,\partial[x/(1-x)]/\partial x = x/(1-x)^2$. Introducing these summations into Eq. (II-4-20) then gives

$$v/v_m = cx/\left[(1-x)(1-x+cx)\right] \qquad \text{(II-4-21)}$$

The Clausius–Clayperon equation may be written as

$$p_0 = (a/b)\exp(-E_L/RT) \qquad \text{(II-4-22)}$$

and since $x = (ap/b)\exp(-E_L/RT)$, it is apparent that $x = p/p_0$ where p_0 is the equilibrium pressure of the pure condensed gas at the temperature in question. Introducing this into Eq. (II-4-21) gives

$$p/v(p_0 - p) = (1/v_mc) + \left[(c-1)/v_mc\right](p/p_0) \qquad \text{(II-4-23)}$$

This is known as the BET equation, and it gives rise to curves such as the lower curve in Fig. II-5. When the curve is plotted according to Eq. (II-4-23), it is linearized as shown by the third column in Table II-4 and c and v_m can be calculated from the slope and the intercept.

Nitrogen is the usual gas employed for gas adsorption methods. A surface coverage of 16 Å$^2 = 16 \times 10^{-16}$ cm^2 per molecule of nitrogen is assumed. Brunauer *et al.* (1959, 1961) points out that in many cases nitrogen gives low results. For instance, with preparations of tobermorite the nitrogen adsorption yielded surface areas that ranged from 20% to 90% of that obtained by water adsorption, and this latter was confirmed by small-angle x-ray scattering.

TABLE II-4

BET Adsorption Data

P (atm)	v (STP) (cm^3)	$p/[v(1-p)]$ calculated	v (return) (STP)
0.2	23.81	0.0105	
0.4	32.52	0.0205	
0.6	49.18	0.0305	63.0
0.8	98.77	0.0405	

Marshall and Sixsmith (1974, 1975) and Hollenbeck *et al.* (1978) have reported similar discrepancies in the case of microcrystalline cellulose.

Example II-4-2 An isotherm is obtained at liquid nitrogen temperature using nitrogen as the adsorbent. The results are listed in Table II-4. Calculate c and the surface area.

Answer Plotting will show that $p/[v(1-p)]$ is linear in p (i.e., the third column in Table II-4 versus the first column). The equation for the line is

$$p/\left[v(1-p)\right] - 0.0205 = \left[(0.0305 - 0.0205)/(0.6 - 0.4)\right](p - 0.4)$$

i.e., $p/v(1-p) = 0.05p + 0.0005$.

It may be seen from the intercept that $1/cv_{\mathrm{m}} = 0.0005$, i.e., $cv_{\mathrm{m}} = 2000$. If c is large, then the slope is $(c-1)/(cv_{\mathrm{m}}) \sim 1/v_{\mathrm{m}} = 0.05$, i.e., $v_{\mathrm{m}} = 20$, so $c = 2000/20 = 100$, justifying the approximation for the slope. $v_{\mathrm{m}} = 20\ \mathrm{cm}^3$ $= (20/22.4)10^{-3}$ mole $= 0.89 \times 10^{-3} \times 6 \times 10^{23} = 5.36 \times 10^{20}$ molecules at $16A^2 = 5.36 \times 10^{20} \times 16 \times 10^{-16} = 85.7 \times 10^4\ \mathrm{cm}^2 = 85.7\ \mathrm{m}^2$.

II-4-3 Pore Distribution

The discrepancy just described may be due to porosity (Marshall *et al.*, 1972). If an isotherm is carried out as shown in Fig. II-5, the values obtained during desorption may differ from those at adsorption. The difference is due to the amount of nitrogen trapped in the pores. This amount will have a lower vapor pressure p_0 than that of the liquid, as given by the Kelvin equation:

$$p_{\mathrm{A}}/p_0 = \exp\left[(-2\gamma V \cos\theta)/RTr_{\mathrm{d}}\right] \qquad \text{(II-4-24)}$$

where γ is surface tension of the adsorbed liquid at temperature T, V its molar volume, θ its wetting angle, r_{d} its capillary dimension, and p_{A} the vapor pressure over the capillary. Hence one may determine at a particular pressure the amount adsorbed during adsorption and desorption. The difference is the amount in the pores. If the pressure is P_1, then the difference corresponds to the volume of the capillary space with a radius equal to or less than r_1, as given by Eq. (II-4-24).

II-4-4 Surface Areas by Permeametry

According to Poisseuille's law,

$$\Delta P/l = \eta 8V/\pi r^4 t \tag{II-4-25}$$

where ΔP is the pressure head, l is the length of the capillary with radius r, η is the viscosity of the liquid, and V/t is the volumetric flow rate in cubic centimeters per second. This latter may be expressed in terms of the linear flow rate v' in the capillary by the following expression:

$$V/t = v'\pi r^2 \quad \text{cm}^3/\text{sec} \tag{II-4-26}$$

Hence one can write

$$\Delta P/l = 8v'\eta/r^2 \tag{II-4-27}$$

When liquid flows through a powder bed with specific surface area (in square centimeters) S_v per cubic centimeter of solid and porosity ϵ, then the so-called hydraulic radius d is given by

$$d = 2r = (1/S_v)[\epsilon/(1-\epsilon)] \tag{II-4-28}$$

The velocity v' of the liquid in the pores is larger than the approach velocity v by a factor of ϵ, i.e., $v' = v/\epsilon$, and this expression and the hydraulic radius may be introduced into Eq. (II-4-27) to give

$$\Delta P/l = 32v\eta(1-\epsilon)^2 S_v^2/\epsilon^3 \tag{II-4-29}$$

In *permeametry* (Fisher subsieve sizer) one introduces an amount of powder corresponding to 1 cm^3 of solid (i.e., weight equal to particle density) into a tube (which is not transparent) of cross section A' (cm^2). The tube is closed by two movable distributor plates. If their distance is l, then the volume is Al and the porosity is assumed to be $(Al-1)/Al$. However if ρ' is such that $\epsilon = 1 - \rho'/\rho_p$ is smaller than $(Al-1)/Al$ then the powder in the tubes will not "fill up" the volume between the two distributor plates. Hence l must be sufficiently small to give reliable values. Conversely, at very small values of l consolidation occurs; i.e., particles fuse together. The curve of d versus ϵ will hence have a plateau or an extremum giving the correct value of S_v (or d).

II-5 DISSOLUTION OF PARTICULATE SYSTEMS

Much has been written in the last two decades about the various theories of dissolution, but whichever theory is used (Dankwerts, 1951; Carstensen, 1977) the basic dissolution equation is always

$$dm/dt = -kA(S - C) \tag{II-5-1}$$

where m is the amount not dissolved, t is time, k is the so-called intrinsic dissolution rate constant, S is saturation concentration, and C is concentration at time t. If C is less than $0.15S$, then to a good approximation one may write (the so-called sink condition equation)

$$dm/dt = -kAS \tag{II-5-2}$$

It is assumed that the particles dissolve in an isotropic fashion, i.e., that the dissolution characteristics of all the crystal faces or all the surfaces are identical. It is also assumed that they are isometric; this term is connected with the shape factor α of the particle, which is given by the expression

$$\alpha = A/V^{2/3} \tag{II-5-3}$$

where A is the surface area of a particle and V is the volume of a particle. That the particle is isometric implies that α is independent of time during dissolution. Cubes and spheres are, e.g., isometric, but parallelepipeds are not. Given N particles of density ρ (g/cm^3), the dissolution equation will take the following form under sink conditions:

$$dm/dt = N\rho(dV/dt) = NAS = Nk\alpha V^{2/3}S \tag{II-5-4}$$

This simplifies to

$$V^{-2/3}\,dV = (k\alpha S/\rho)\,dt \tag{II-5-5}$$

which can be integrated to give

$$V_0^{1/3} - V^{1/3} = (\alpha kS/3\rho)t \tag{II-5-6}$$

Introducing now the nomenclature

$$K = kS/\rho \tag{II-5-7}$$

and multiplying through by $(M_0/V_0)^{1/3}$, where M denotes mass undissolved at time t and where the subscript zero implies initial conditions, then gives

$$M_0^{1/3} - M^{1/3} = \left[(M_0/V_0)^{1/3}K\alpha/3\right]t \tag{II-5-8}$$

This is (a slightly modified form of) the Hixson–Crowell equation (Hixson and Crowell, 1931) and has been verified by many authors (e.g., Pothisiri and Carstensen, 1975). A more complicated form can be derived for nonsink conditions (Short *et al.*, 1972; Patel and Carstensen, 1975).

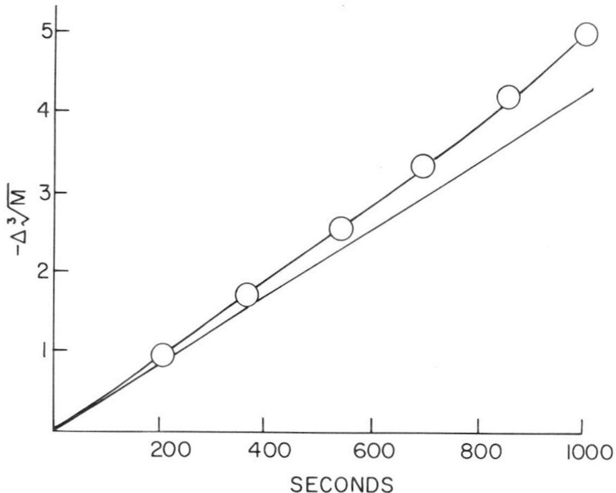

Fig. II-7 Cube root dissolution plots of oxalic acid tablets with $a = 2.75$.

Deviations from the cube root law (in monodisperse systems) have been reported in literature and are mostly characterized by a slight upward curvature (Goyan, 1965). It should be emphasized that these upward curvatures are never large, and this is exemplified in Fig. II-7. In other words the Hixson–Crowell law is a good working model. Theories have been put forth in the past to explain this upward curvature, but the fact that the lack of isometry of most real particles can explain the upwards curvatures has never been voiced. If it is taken into account that the shape factor α is time dependent, i.e., of the form $\alpha(t)$, then one can by some means calculate the integrated mean value of α at any time t', i.e.,

$$\alpha_{\mathrm{m}} = \frac{1}{t'} \int_0^{t'} \alpha(t)\, dt \tag{II-5-9}$$

The method to be described in what follows does this by introducing a reduced time parameter for the dissolution of a cylinder with initial diameter d_0 and initial height h_0:

$$u = 2(kS/h_0\rho)t \tag{II-5-10}$$

Linearity of u with t implies constancy of K [Eq. (II-5-7)] at any time t, implying adherence to Eq. (II-5-8), and an example of this is shown in Fig. II-8.

The reason for selecting a cylinder is that when the ratio $a = d_0/h_0 > 1$ becomes large, the shape of the cylinder will approach that of a plate, whereas when the value of the ratio $b = h_0/d_0 > 1$ becomes large, the shape of the cylinder will approach that of a needle. It can be shown

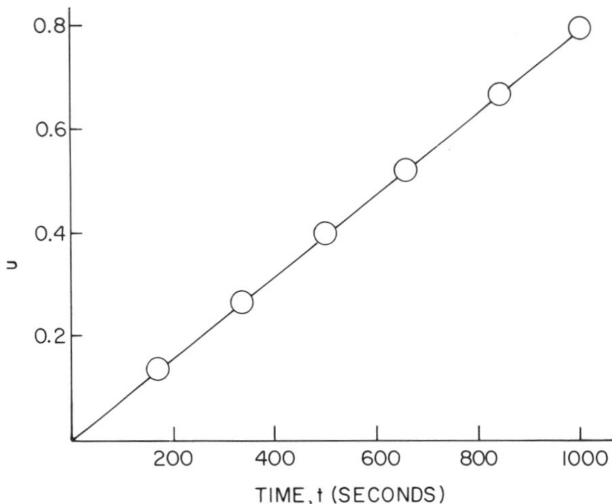

Fig. II-8 Data from Fig. II-7 plotted according to Eq. (II-5-10) where u is determined via Eq. (II-5-11).

geometrically that the percent of material dissolved at time t is

$$p = 100\left[1 - (1 - u/a)^2(1 - u)\right] \qquad \text{(II-5-11)}$$

At any time point, therefore, it is possible to calculate u from Eq. (II-5-11), which is a cubical equation. There will be two imaginary and one real root, and the latter, of course, applies. An example of the linearity obtained in this fashion is shown in Fig. II-8. The diameter and the height can, of course, be expressed in terms of h_0, d_0, and time (and hence u), so

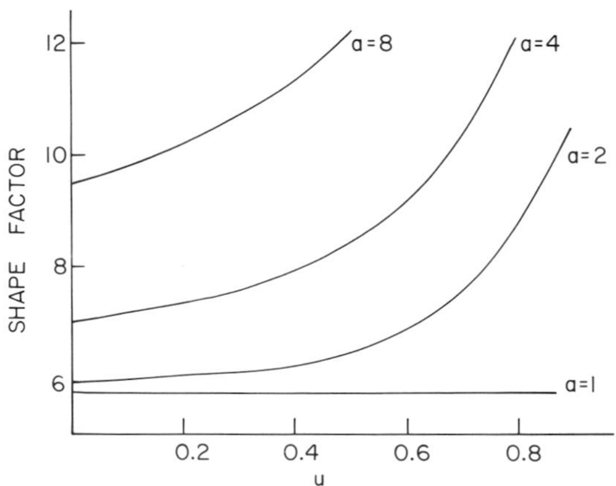

Fig. II-9 α_m versus u for various a values.

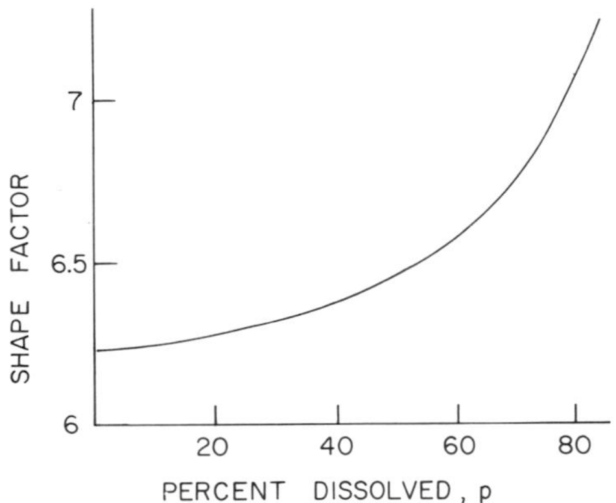

Fig. II-10 α_m versus p $(a = 2.75)$.

that the shape factor can be calculated from Eq. (II-5-3) and estimated via the equivalent of Eq. (II-5-9):

$$\alpha_m = (1/u')\int_0^{u'} \alpha(u)\, du \tag{II-5-12}$$

and the following expression results:

$$\alpha_m = 3.69\left\{ \tfrac{1}{2}\left[(a-u)/(1-u)\right]^{2/3} + \left[(1-u)/(a-u)\right]^{1/3} \right\} \tag{II-5-13}$$

A plot of α_m versus u according to Eq. (II-5-13) is shown in Fig. II-9 and a plot of α versus p is shown in Fig. II-10.

Because of general precision considerations, if a 5% change in α occurred and were neglected, then a serious error would not have been committed (and would in most cases be within analytical detection). In Fig. II-10 for instance, $\alpha = 6.25$ initially, so that if α rose to a value of $1.05 \times 6.25 = 6.56$, then it would still be permissible to consider α constant. Note from Fig. II-10 that this occurs at $p = 60\%$; i.e., for $a = 2.75$ the Hixson–Crowell equation can be applied for the format shown considering α constant. The value of 60% dissolved which was calculated will be denoted the 5% limit number L, and a graph of L versus a is shown in Fig. II-11a. Figure II-11b shows a similar graph for needles, where the ratio is $h = h_0/d_0$. It may be seen from these curves that isometry, although theoretically of significance, does not play a *large* role in dissolution of particles. A series of publications by Pedersen and Brown (1975, 1976) have provided examples of this for more complicated geometric shapes.

One can determine k as a function of the linear velocity $\bar{u}$ of liquid passing by it in column dissolution experiments of monodisperse sieve cuts, and this has been done for oxalic acid (Carstensen and Dhupar,

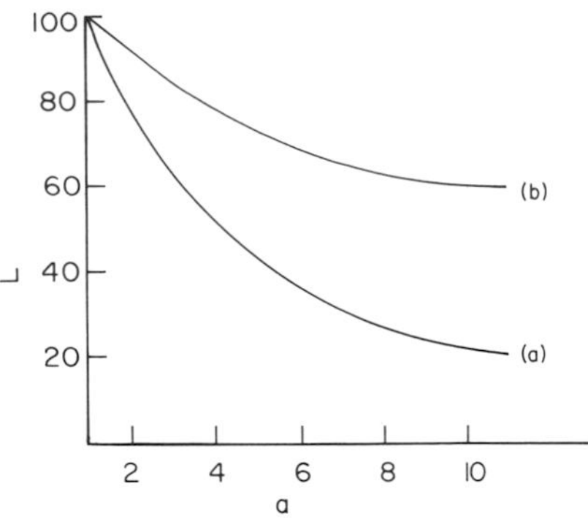

Fig. II-11 (a) Values of the 5% limit number L as a function of a. (b) Values of the 5% limit number L (percent dissolved) as a function of b.

1976). At low velocities the dissolution rate constant was found to be linear in $\bar{u}$. Lai and Carstensen (1978) found the data at higher values of $\bar{u}$ to be parabolic and to fit the equation

$$\ln k = \left[(0.59 \pm 0.27)\ln(\bar{u})\right] - (7.38 \pm 0.65) \qquad \text{(II-5-14)}$$

The slope is therefore of the order of $\frac{1}{2}$, as in hydrodynamic theory, which predicts that the hydrodynamic layer (i.e., the layer in which the velocities decrease from $\bar{u}$ in the bulk to zero at the solid surface) has a thickness δ given by

$$\delta = q(\bar{u})^{-n} \qquad \text{(II-5-15)}$$

where q is a proportionality constant and where n has values between $\frac{1}{2}$ and 1. But this layer is not (necessarily) the same as the concentration gradient layer (which is h cm thick).

In the case of polydisperse powder, Carstensen and Musa (1972, p. 273) based on initial work done by W. Higuchi and Hiestand (1963), suggested from data obtained by computer simulation that a cube root law would also hold for log-normally distributed powders and have the form (σ denoting standard deviation)

$$1 - (M/M_0)^{1/3} = \left[2kS/r_0\rho f(\sigma)\right]t = (Q/r_0) \qquad \text{(II-5-16)}$$

where r_0 is the original (geometric) mean diameter and $f(\sigma)$ is a function of the standard deviation of the log-normal distribution. Brooke (1975) derived the relation

$$f(\sigma) = e^{-c\sigma^2} \qquad \text{(II-5-17)}$$

a point which was experimentally verified by Carstensen and Patel (1975). The dissolution is obviously biphasic (since it changes profile at the critical time when the smallest particle disappears completely), and this point has also been demonstrated experimentally.

Carstensen (1977) treated the data by Prescott *et al.* (1970) of plasma levels of phenacitin of different diameter powders using Eq. (II-5-16). If a certain length of time t' is available as a "window" during which absorption takes place (and if parameter values do not change during t') then the fraction absorbed F is related to the amount M which is undissolved at time t' by the relation $F = 1 - M/M_0$, so that Eq. (II-5-16) becomes

$$(1 - F)^{1/3} = 1 - Q/r_0 \tag{II-5-18}$$

The in vivo data by Prescott fit such a curve with a negative slope $-Q$ and an intercept of unity [as predicted by Eq. (II-5-18)].

II-6 STABILITY OF PARTICULATE SOLIDS

The stability of pure solids differs from that of solutions in that it is often dictated by physical processes. There are two main types: (a) topochemical reactions and (b) surface-initiated reactions.

II-6-1 Contracting Geometries

Langmuir (1916) was the first to propose that there was a connection between geometrical aspects of a solid and its decomposition, and he hypothesized that the reactions occurred and started at surfaces and boundaries of the solid. Jacobs and Tompkins (1955) outlined in a review the three predominant cases.

The contracting cylinder (Fig. II-12a) is one for which one assumes that the intact chemical substance has an initial radius r_0 which decreases linearly with time, i.e.,

$$r = r_0 - k_1 t \tag{II-6-1}$$

At time t, there will be unchanged chemical substance A inside the cylinder of radius r, and decomposition product B outside (Fig. II-12b). One can therefore calculate the fraction decomposed as

$$x = hn\rho\pi\left(r_0^2 - r^2\right)/hn\rho\pi r_0^2 \tag{II-6-2}$$

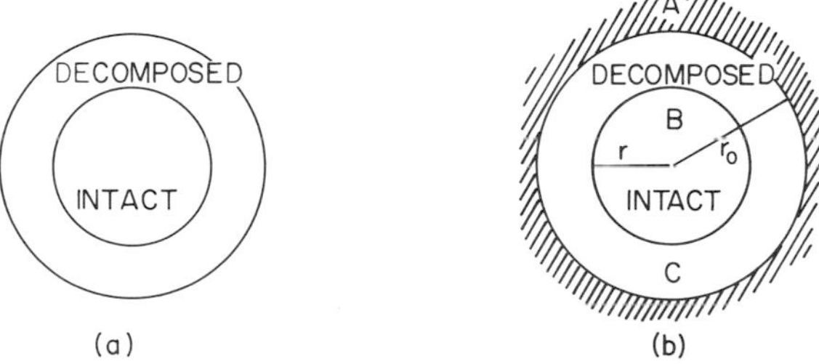

Fig. II-12 (a) Contracting sphere or cylinder. (b) Reaction of A with spherical B through a layer of decomposition product.

Here n is the number of particles and h is the height of each cylindrical particle. Note that n and h cancel out, so that the amount that is retained can be obtained from Eqs. (II-6-1) and (II-6-2):

$$1 - x = 1 + (k_1/r_0)^2 t^2 - (2k_1/r_0)t \qquad \text{(II-6-3)}$$

or

$$\sqrt{(1 - x)} = 1 - (k_1/r_0)t \qquad \text{(II-6-4)}$$

Figure II-12a could be considered to represent a sphere, and in this case one can similarly obtain

$$x = \frac{\frac{4}{3}\pi r_0^3 - \frac{4}{3}\pi(r_0 - k_3 t)^3}{\frac{4}{3}\pi r_0^3} = 1 - \left[\frac{1 - k_3 t}{r_0}\right]^3 \qquad \text{(II-6-5)}$$

This may be written as

$$(1 - x)^{1/3} = 1 - (k_3/r_0)t \qquad \text{(II-6-6)}$$

One can obviously obtain a similar type of equation if one starts with a cubical geometry, and Eq. (II-6-6) is often called the contracting cube equation.

Roginski and Schultz (1928) studied the decomposition of potassium permanganate and found Eq. (II-6-6) to hold. Jander (1927) derived the following equations from geometrical principles as well:

$$(1 - x)^{1/2} = 1 - \Gamma t \qquad \text{(II-6-3}')$$

$$(1 - x)^{1/3} = 1 - \Gamma t \qquad \text{(II-6-6}')$$

$$(1 - x)^{1/3} = 1 - (\Gamma t)^{1/2} \qquad \text{(II-6-7)}$$

The derivation of Eq. (II-6-7) is instructive. Consider a small amount of material B in the form of spheres embedded in matrices of substance A, and consider that the decomposition reaction is $A + B \rightarrow C$. This is depicted in Fig. II-12b. At time t, A will have to diffuse through a layer of decomposition product C, the thickness of which is $r_0 - r$. (It is assumed here that A is the species which must diffuse; i.e., A reacts rapidly with B once it has arrived to the B layer by diffusing through the decomposition layer.) As A is lost it causes a proportional increase in the decomposition layer by a factor which will be denoted γ, and hence

$$-dA/dt = \gamma(dl/dt) = q/l \qquad \text{(II-6-8)}$$

Here l equals $r_0 - r$ and is the width of the decomposition layer, and q is a type diffusional coefficient. The last step in Eq. (II-6-8) is obtained via Fick's law. This latter equation may now be integrated to

$$l^2 = (2q/\gamma)t = k't$$

or

$$r_0 - r = \sqrt{k' t} \tag{II-6-9}$$

Note that

$$1 - x = \pi r^3 / \pi r_0^3 = (1 - \sqrt{k'' t}\,)^3 \tag{II-6-10}$$

The last step was accomplished by means of Eq. (II-6-9). In Eq. (II-6-10)

$$k'' = k' / r_0^2 = 2q / \gamma r_0^2$$

Equation (II-6-10) can be written as

$$\left[1 - (1 - x)^{1/3}\right]^2 = k'' t$$

which is equivalent to Eq. (II-6-7).

Example II-6-1 Nelson *et al.* (1974) found the following salicylic acid contents in an aspirin preparation at 40°C: 3.08% after one day and 4.40% after two days. Do these data follow Eq. (II-6-7)?

Answer For $t = 1$, $\sqrt{t} = 1$ and $1 - (1 - x)^{1/3} = 0.0104$; i.e., the slope is 0.0104.

For $t = 2$, $\sqrt{t} = 1.41$ and $1 - (1 - x)^{1/3} = 0.0149$; i.e., the slope is $0.0149/1.41 = 0.0106$, in good agreement. In other words, the data follow Eq. (II-6-7).

A set of curves of the types described in Eqs. (II-6-3′), (II-6-6′), and (II-6-7) is drawn in Fig. II-13a and tabulated in Table II-5. We shall see that many decompositions in solid dosage forms and solids are approximated by first-order kinetics, and therefore the above have been compared with a graph of an equation of the type

$$1 - x = e^{-kt} \tag{II-6-11}$$

This is obviously a first-order case. All the curves shown in Fig. II-13a have been fitted through the point $\{3, 0.28\}$. When γt is small, Eqs. (II-6-3′) and (II-6-6′) become

$$1 - x \sim 1 - 2k_1 t / r_0 \tag{II-6-3″}$$

$$1 - x \sim 1 - 3k_3 t / r_0 \tag{II-6-6″}$$

These may be compared to an expansion of Eq. (II-6-11). For small values of kt, this becomes

$$1 - x = 1 - kt \tag{II-6-11a}$$

These are obviously comparable if $k = 3kt/r_0$ or $k = 2k_1/r_0$. This comparison is shown in Table II-5 and in Fig. II-13b. There are a few pharmaceutical examples of the utility of Eq. (II-6-7), such as the work of Horikoshi and Himuro (1966) and of Nelson *et al.* (1974).

TABLE II-5

Decompositions According to Eqs. (II-6-3′), (II-6-6), (II-6-6′), and (II-6-11)

Time, t	$\gamma = 0.05\ \text{hr}^{-1}$, Eq. (II-6-3′)	$\gamma = 0.035\ \text{hr}^{-1}$, Eq. (II-6-6)	$\gamma = 0.00359\ \text{hr}^{-1}$, (Eq. II-6-7)	$\gamma = 0.11\ \text{hr}^{-1}$, Eq. (II-6-11)	k_3t/r_0, Eq. (II-6-6)	$1 - x$, Eq. (II-6-6)	$\ln(1 - x)$
1	0.1	0.1	0.17	0.10	0.01	0.970	− 0.030
2	0.19	0.20	0.23	0.20	0.02	0.941	− 0.061
3	0.28	0.28	0.30	0.28	0.04	0.885	− 0.122
5	0.44	0.44	0.35	0.42	0.06	0.830	− 0.186
10	0.75	0.73	0.47	0.67	0.10	0.729	− 0.316
15	0.94	0.89	0.55	0.81	0.15	0.614	− 0.488
20			0.61		0.20	0.512	− 0.669
25			0.66				

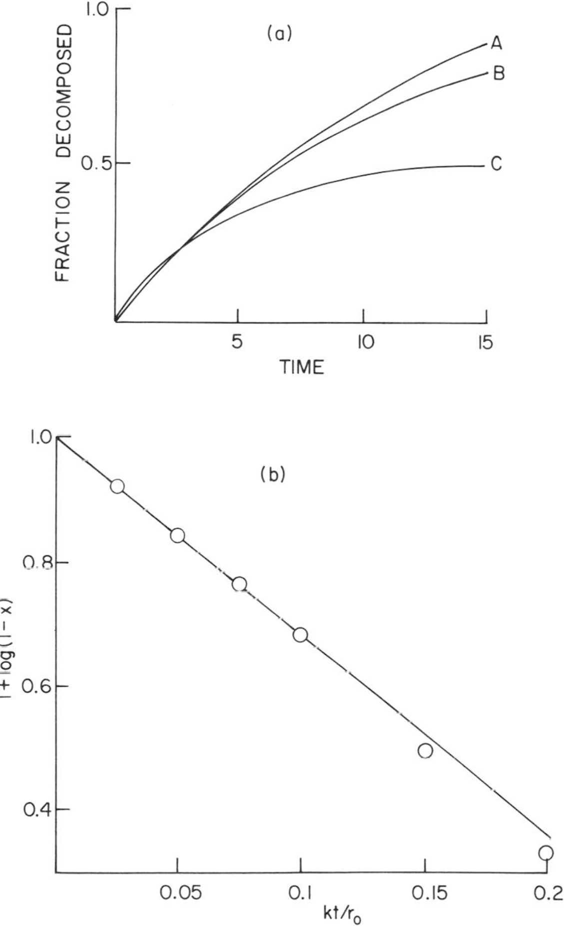

Fig. II-13 (a) Decomposition data from Table II-5. A is a contracting cube, B is product layer diffusion controlled, and C is first-order decomposition. (b) Data following Eq. (II-6-6) plotted as first order.

II-6-2 Surface Initiated Reactions

These arise from the following types of reactions:

1. solid → solid + gas (PASA)
2. solid → solid + solid
3. solid → solid + liquid
4. solid → liquid + gas (PABA)
5. solid → gas + gas (NH_4NO_2)

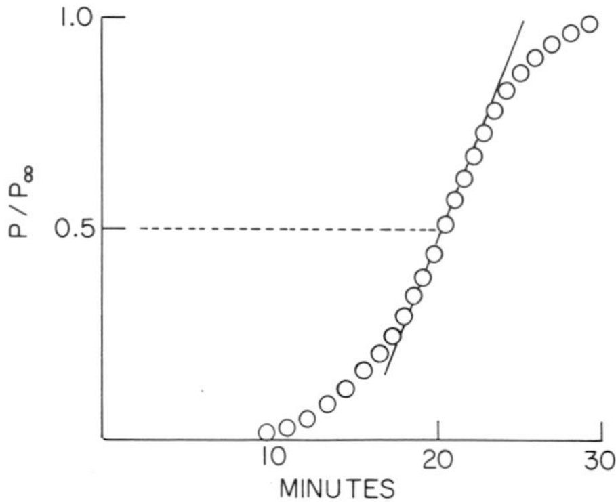

Fig. II-14 Pressure evolved P as a function of time in thermal decomposition of silver permanganate. [After Goldstein and Flanagan (1964). Adapted with permission of the copyright owner.]

When one follows the carbon dioxide evolution of types 1 and 4, one obtains sigmoid curves when the mole fraction decomposed (x) is plotted as the ordinate with time as the abscissa. Frequently, as shown in Fig. II-14, there is initially an induction period ($0 < x < 0.1$). Next there is an acceleratory period ($0.1 < x < 0.5$). Finally there is a decay period. This latter is not quite as reproducible as the two first periods. The inflection point frequently happens at $x = P/P_{\infty} = 0.5$, where P is the pressure of the gas evolved. Prout and Tompkins (1944) have put forward the model most generally accepted as explaining how this type decomposition occurs. It is assumed that there is first formation and growth of active nuclei on the surface of the substance. These type of nuclei and their growth have been demonstrated by several investigators (Prout and Tompkins, 1944; Kohlschütter, 1927; Coppeck *et al.*, 1931; Garner and Southon, 1935; Garner and Pike, 1937; Coppet and Garner, 1936). They usually occur on the surface but can also occur inside (Coppet and Garner, 1936). In the model it is assumed that the reaction probability according to scheme 1, 4, or 5 is proportional to the number of nuclei. The nuclei are assumed to propagate with a probability of α and to terminate with a probability of β. One assumes now that the initial number of potential nucleus formation sites is N_0 and that at time t there will be N nuclei present, so therefore

$$dN/dt = k(N_0 - N) \tag{II-6-12}$$

At time point t there hence are $N_0 - N$ sites still present for use. If at that particular point one disregards the termination possibility, then this can be integrated to read

$$N = N_0[1 - \exp(-kt)] \sim N_0 kt \tag{II-6-13}$$

The last step, the approximation, is valid for small values of kt. Equation

(II-6-13) is now differentiated to give

$$dN/dt = kN_0 \tag{II-6-14}$$

However, a nucleus has a probability α of branching, and one may therefore write

$$dN/dt = kN_0 + \alpha N \tag{II-6-15}$$

After a while branches will start merging (and the probability of this is β), so that Eq. (II-6-15) must include this point as well. It is reasonable to assume that the initial potential nucleation sites are rapidly used up so that Eq. (II-6-15) can be written as

$$dN/dt = (\alpha - \beta)N \tag{II-6-16}$$

Note that α and β are both functions of time, i.e., functions of t and x; it is therefore not possible to integrate Eq. (II-6-16) directly. At $x = 0.5$ (or at the inflection point), $\alpha = \beta$, and it is assumed that $\alpha = 1$ and $\beta = 0$ at $t = 0$ ($x = 0$) and hence the function $\beta = 2x\alpha$ can reasonably be used as a trial function for the model. One can therefore write

$$dN/dt = \alpha(1 - 2x)N \tag{II-6-17}$$

It is now assumed that the decomposition rate is proportional to N, the number of nuclei; i.e.,

$$dx/dt = k_1 N \tag{II-6-18}$$

This can be introduced into Eq. (II-6-17) to give

$$\frac{dN}{dt} = \frac{dN}{dx}\frac{dx}{dt} = \frac{\alpha(1-2x)}{k_1}\frac{dx}{dt} \tag{II-6-19}$$

In other words,

$$N = (\alpha/k_1)(x - x^2) \tag{II-6-20}$$

Under the assumptions that $N_0 \ll N$ and that $N = 0$ implies $x = 0$, Eq. (II-6-20) is now introduced into Eq. (II-6-18) to give

$$\ln[x/(1-x)] = \gamma t + C \tag{II-6-21}$$

Example II-6-2 The data in Table II-6 were found for the decomposi-

TABLE II-6

Decomposition of 2 mmole of Solid

Time	Gas (mmole)	x	$\ln[x/(1-x)]$
1	0	0	
2	0.5	0.18	− 1.715
3	1.2	0.3	− 0.847
4	2.0	0.5	0
5	2.8	0.7	+ 0.847

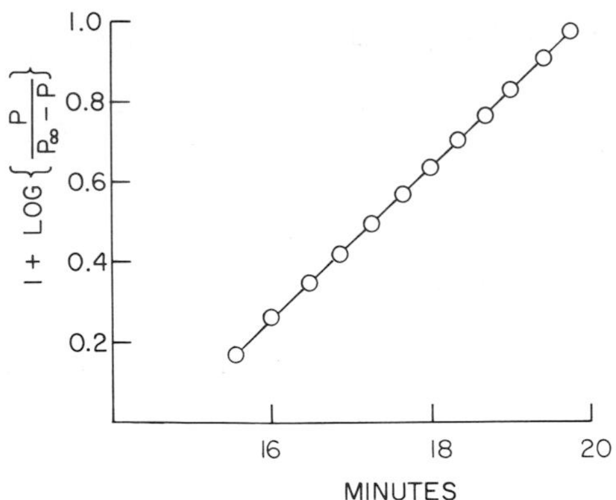

Fig. II-15 Data from Fig. II-14 plotted according to Eq. (II-6-21). [After Goldstein and Flanagan (1964). Adapted with permission of the copyright owner.]

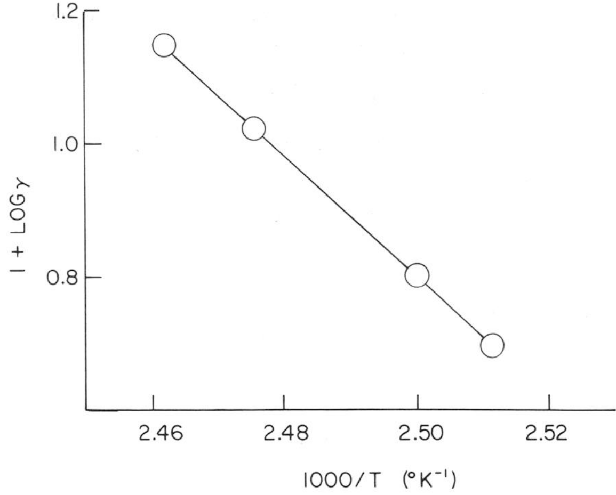

Fig. II-16 Data of the type in Fig. II-15 plotted by Arrhenius plotting. [After Goldstein and Flanagan (1964). Adapted with permission of the copyright owner.]

tion of a medicinal which is a substituted hydrazine ($RNH—NH_2 \rightarrow RH + N_2 + H_2$). Do they follow Eq. (II-6-21)?

Answer The quantities x and $\ln[x/(1-x)]$ are listed in Table II-6. The slope of the line is (2–3 hr) $1.715 - 0.847 = 0.868\ \text{hr}^{-1}$, which is in good agreement with 0.847 for the points that follow. The equation is then

$$\ln[x/(1-x)] + 0 = 0.868(t-4)$$

Data from Fig. II-14 are plotted according to Eq. (II-6-21) in Fig. II-15. The slopes, as seen in Fig. II-16, adhere well to an Arrhenius equation.

II-6-3 Modified Prout–Tompkins Equation

Although the Prout–Tompkins equation [Eq. (II-6-21)] frequently applies, there are many systems where the inflection point occurs at x considerably smaller than $\frac{1}{2}$. In such cases (Fig. II-17) the decompositions will look predominantly first order. To explain this the following revision of the Prout–Tompkins theory is made: it is assumed that the nucleation starts at $t = t_0$ (> 0) (Carstensen and Pothisiri, 1975) and that the probability of propagation is α_0 (< 1); it is then also assumed that $\alpha = \beta$ at $t = t_i$ where t_i is the time at which the inflection takes place. At infinite time the probability of termination is denoted β_0. One can now find a function

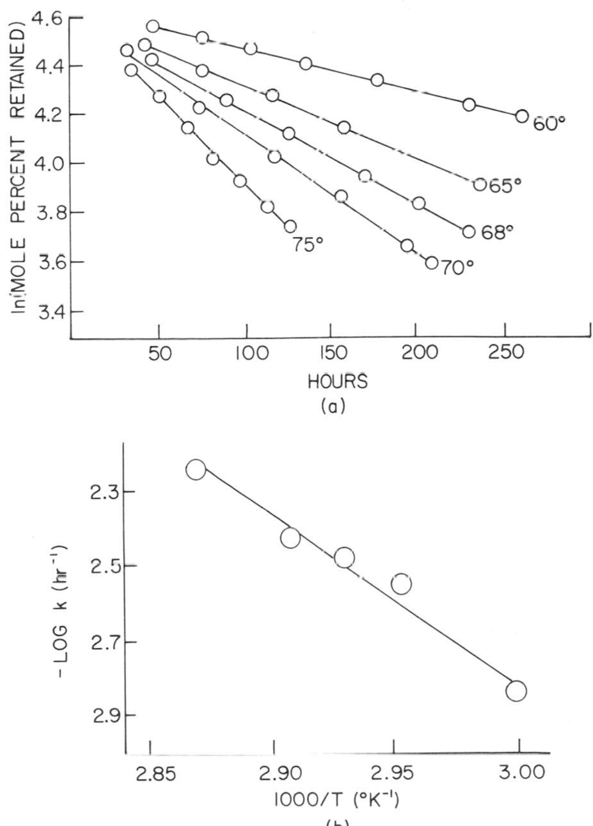

Fig. II-17 (a) Decomposition of *p*-aminosalicylic acid as a function of time and temperature. [After Carstensen and Pothisiri (1975). Reproduced with permission of the copyright owner, the *Journal of Pharmaceutical Science*.] (b) Arrhenius plot of data in Fig. II-17a.

which adheres to these boundary requirements, and one such function is

$$\alpha - \beta = [(\alpha_0 + \beta_0)(t_0 - q)/(t - q)] - \beta_0 \tag{II-6-22}$$

where

$$q = [(\alpha_0 + \beta_0)t_0 - \beta_0 t_i]/\alpha_0 \tag{II-6-23}$$

Equation (II-6-23) may be arranged as $\alpha_0(q - t_0) = \beta_0(t_0 - t_i)$, $t_0 - t_i < 0$, and it therefore follows that the quantity $q - t_0 < 0$, i.e., $t_0 > q > 0$. Eq. (II-6-23) can be inserted into Eq. (II-6-16) to yield

$$dN/N = \{[(\alpha_0 + \beta_0)(t_0 - q)/(t - q)] - \beta_0\}\, dt \tag{II-6-24}$$

This may be integrated to give

$$\ln N = (\alpha_0 + \beta_0)(t_0 - q)\ln(t - q) - \beta_0 t + \ln Q \tag{II-6-25}$$

In this $\ln Q$ is an integration constant. This equation may be written as

$$N = Q(t - q)^{(\alpha_0 + \beta_0)(t_0 - q)} \exp(-\beta_0 t) \tag{II-6-26}$$

This can be inserted into Eq. (II-6-18) to give

$$dx/dt = k'Q(t - q)^{(\alpha_0 + \beta_0)(t_0 - q)} \exp(-\beta_0 t) \tag{II-6-27}$$

In the case in which the quantity $t_0 - q$ is negligibly small, this latter equation may be written as

$$dx/dt = k'Q \exp(-\beta_0 t) \tag{II-6-28}$$

This can be integrated to give

$$\begin{aligned} x &= 1 - \exp[-\beta_0(t - t_0)] &\quad \text{when} \quad t \geqslant t_0 \\ x &= 0 &\quad \text{when} \quad t < t_0 \end{aligned} \tag{II-6-29}$$

Note that initial conditions for $x = 0$ at $t < t_0$ have been used in proceeding from Eq. (II-6-28) to Eq. (II-6-29), and also that the boundary condition $x = 1$ when $t \to \infty$ has been used. The latter boundary condition implies that $k'Q/\beta_0 = 1$. Note that Eq. (II-6-29) is the equation for a first-order reaction. When t_0 is not close to the value of q, then as shown below the expressions will become more complex. If in Eq. (II-6-27) we make the substitutions

$$(\alpha_0 + \beta_0)(t_0 - q) = m, \qquad u = t - q \quad (du = dt, t = u + q) \tag{II-6-30}$$

and then introduce

$$z = \beta_0 u \qquad (dz = \beta_0\, du) \tag{II-6-31}$$

we obtain

$$dx = [k'Q \exp(-\beta_0 q)\beta_0^{-(m+1)}] z^m \exp(-z)\, dz = M z^m \exp(-z)\, dz \tag{II-6-32}$$

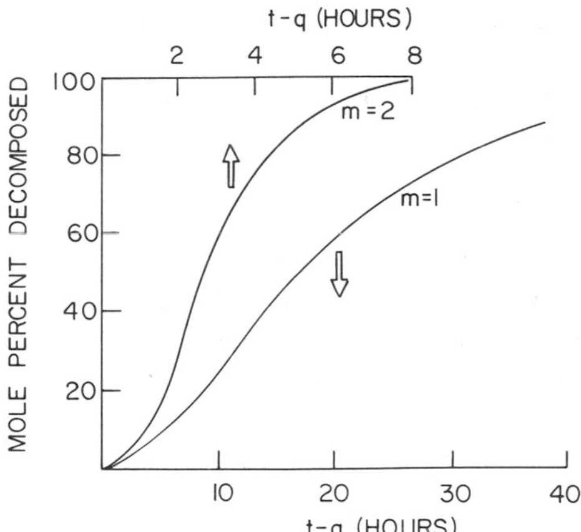

Fig. II-18 Theoretical decomposition curves according to Eq. (II-6-36) with $m = 1$ and $m = 2$ and $\beta_0 = 0.1\ \text{hr}^{-1}$.

Here M is equal to

$$M = k'Q \exp(-\beta_0 q)\,\beta_0^{-(m+1)} \tag{II-6-33}$$

If Eq. (II-6-32) is integrated with $t = t_0$ as lower limit and ∞ as upper limit, the result should be $x = 1$, since this corresponds to total decomposition. Since $t_0 > q$ the integration actually yields a value of unity if it is carried out from q to infinity. At $t = q$, $u = z = 0$, so the integration of Eq. (II-6-32) from $z = 0$ to infinity gives

$$1 = M\int_0^\infty z^m \exp(-z)\,dz = \Gamma(m+1) \tag{II-6-34}$$

This can be arranged to give

$$M = \Gamma^{-1}(m+1) \tag{II-6-35}$$

where $\Gamma(m+1)$ is the Γ function evaluated at $m + 1$. Equation (II-6-32) can be shown (by integrating by parts) to give a general solution of

$$x = 1 - \left\{\exp[-\beta_0(t-q)]\sum_0^m [\beta_0(t-q)]^{m-1}\frac{m!}{(m-1)!}\right\} \tag{II-6-36}$$

where m is an integer. Various examples of this type of curve are shown in Fig. II-18. Observe that the inflection point of these curves will occur when

$$d^2x/dz^2 = 0 = M\exp(-z)\,\{z^{(m-1)}\}(m-z) \tag{II-6-37}$$

i.e., when $z = \beta_0(t-q) = m$.

II-6-4 Systems with Liquid Decomposition Product Layers

The most common reaction is not scheme 1 but rather scheme 4 (Section II-6-2). Carstensen and Musa (1972b) have shown that this type of reaction occurs for a series of substituted benzoic acids, and Carstensen and Pothisiri (1975) have shown it to happen for many substituted salicylic acids as well. Aminobenzoic acid, for instance, decomposes into aniline and carbon dioxide. During this the aminobenzoic acid will dissolve in the aniline which is formed and the extent to which it dissolves, in mole/mole of aniline (the solubility), is S. At time t, at which point an amount x (in moles) is decomposed, there will be therefore a number of moles Sx of aminobenzoic acid in solution, whereas a number $1 - x - Sx$ of moles of aminobenzoic acid is still in the solid state. With k_s and k_a the first-order rate constants for solid and dissolved state, respectively, one can therefore write the rate equation as

$$dx/dt = k_s(1 - x - Sx) + k_a Sx = k_s + \Gamma x \tag{II-6-33a}$$

The term Γ is given by

$$\Gamma = -(k_s + k_s S - k_a S) \tag{II-6-34a}$$

This may be integrated and yields

$$\ln(1 + \Gamma x/k_s) = \Gamma t \tag{II-6-35a}$$

In this manner one can find Γ/k_s by using it as an adjustable parameter. Of course, Γ is known from the slope, so that k_s can be found by dividing the slope by the adjustable parameter. By observing at what point liquefaction occurs S is determined directly. After this the aminobenzoic acid will simply decompose by a first-order reaction in solution, and one can find k_a from this part of the curve as well as from Eq. (II-6-34a), since here S and k_s are by now known. It has been shown by Carstensen and Musa (1972, p. 1112) and by Pothisiri and Carstensen (1975) that k values found by the two methods give comparable results. The curves observed in these types of systems will also be sigmoid, since the decomposition curve in that part of the curve up to the point where liquefaction occurs is upward convex and given by the equation

$$x = (k_s/\Gamma)[\exp(\Gamma t) - 1] \tag{II-6-36a}$$

After liquefaction the curve becomes convex in the downward direction and follows the equation

$$(1 - x)/(1 - x') = \exp[-k_a(t - t')] \tag{II-6-37a}$$

where $x' = 1/(1 + S)$ at $t = t'$. The k_a and k_s values obtained by Carsten-

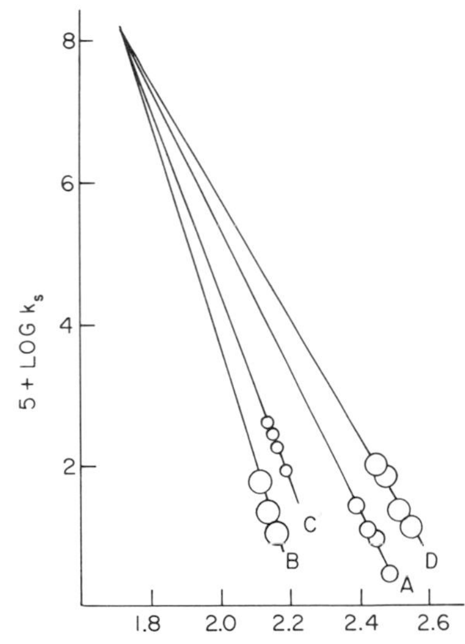

Fig. II-19 Plot showing possible isokinetic points in the solid state decomposition of *p*-substituted benzoic acids. Key: A, amino; B, dimethylamino; C, hydroxy; and D, methyl-amino. [After Carstensen and Musa (1972, p. 112). Reproduced with permission of the copyright owner, the *Journal of Pharmaceutical Science*.]

sen and Musa (1972, p. 1112) can be plotted on Arrhenius plots. Note that in Fig. II-19 the k_s curves seem to intersect at one point and there is the possibility of such an isokinetic point. Guillory and Higuchi (1962) in studying a series of vitamin A derivatives suggested that rate constants k_A could be related to melting point T_m of the compounds in the form of the

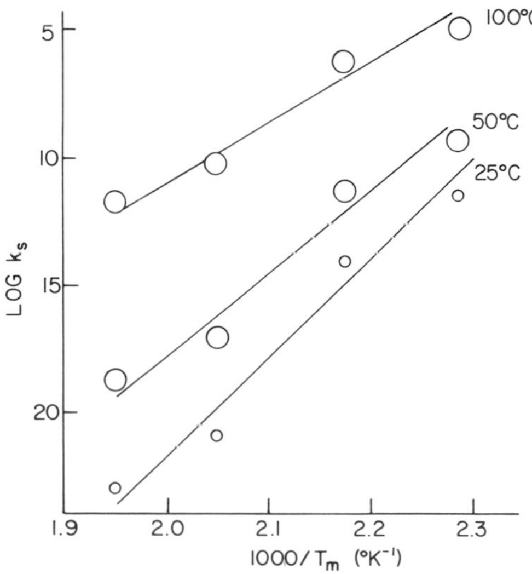

Fig. II-20 Decomposition rate constant in solid state (k_s) as a function of melting point T_m of substituted benzoic acids. [After Carstensen and Musa (1972, p. 1112). Reproduced with permission of the copyright owner, the *Journal of Pharmaceutical Science*.]

equation

$$\log k_A = (\theta / T_m) + \sigma \qquad \text{(II-6-38)}$$

Carstensen and Musa (1972, p. 1112) found in the series of benzoic acid studies that the rate constants followed the Guillory–Higuchi equation fairly well (Fig. II-20).

II-6-5 Linear Free Energy Relations

Pothisiri and Carstensen (1975) found that the substituted benzoic acid series and salicylic acid series that had been investigated followed Hammett-type plots (Fig. II-21). The slopes, however (the reaction parameters), were of opposite signs for the two series. By x-ray crystallography they determined the molecular structures of the crystals and found that the compounds were dimeric, that is, the molecules occurred in pairs in the crystal lattice. In the case of substituted benzoic acids there is a close proximity between the substituant X of one dimer and one carboxyl group in the neighboring dimer, so that the decomposition is intermolecular. In the case of salicylic acid, however, this is not the case, so that in this case the decomposition is intramolecular. In plotting the solid and liquid rate constants as a function of temperature good Arrhenius relations were found. If these lines are extrapolated to high temperatures they will eventually cross and there will be a temperature at which the two rate constants are equal. This temperature is far above the melting point of the

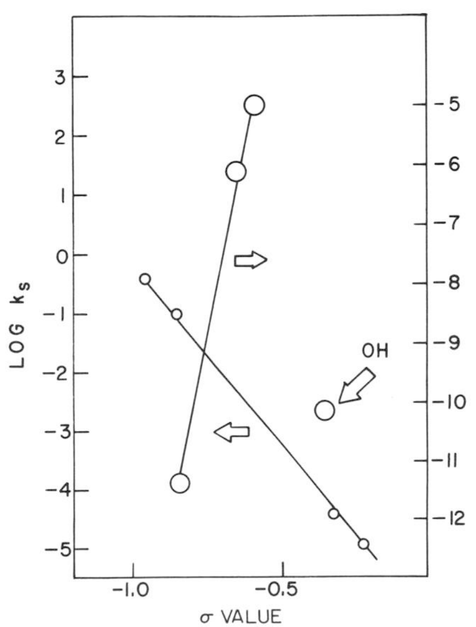

Fig. II-21 Hammett-type plot of a series of *p*-substituted salicylic acids (left ordinate, small circles) and *p*-substituted benzoic acids (right ordinate, large circles). [After Pothisiri and Carstensen (1975). Reproduced with permission of the copyright owner, the *Journal of Pharmaceutical Science*.]

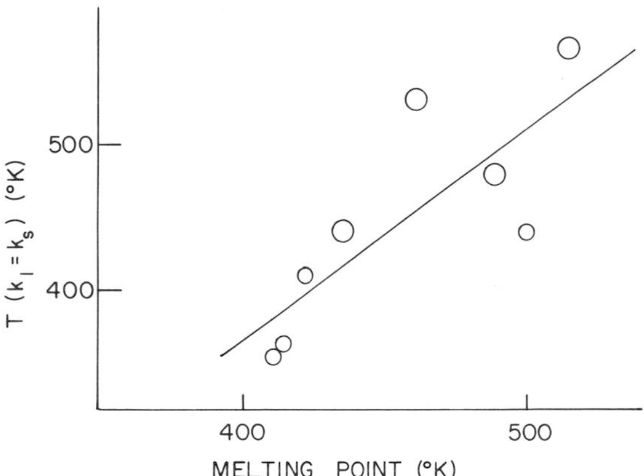

Fig. II-22 Temperature at which the solid and liquid decomposition rate constants are equal as a function of melting temperature for a series of substituted salicylic acids. [After Pothisiri and Carstensen (1975). Reproduced with permission of the copyright owner, the *Journal of Pharmaceutical Science*.]

compound. It is interesting to note that the salicylic acid series does not follow the Higuchi equation (II-6-38), but it may be seen from Fig. II-22 that when the temperature at which the liquid and solid rate constants are equal is plotted as a function of the melting point a linear correlation is obtained.

II-7 APPARENT DENSITY AND BED POROSITY

If a powder is poured (cascaded) into a container, it forms a *bed*, part of which is solid, part of which is void space (air). In work with solids there is the following conventional density terminology:

True density (of the *material*) ρ: the mass of 1 cm^3 of nonporous crystal.

Particle density ρ_p: the mass of a solid particle in grams per cubic centimeter of particle, including particulate void space.

Apparent density ρ': the bed density, i.e., the mass of 1 cm^3 of particles including (interparticulate) void spaces between particles as well as (intraparticulate) void space within the particles.

The total porosity ϵ is the sum of the interparticulate porosity ϵ_b (b for bed) and intraparticulate porosity ϵ_p (p for particle). It follows that

$$\epsilon = 1 - \rho'/\rho \tag{II-7-1}$$

$$\epsilon_p = 1 - \rho_p/\rho \qquad \text{so} \qquad \epsilon_b = \epsilon - \epsilon_p = (\rho_p - \rho')/\rho \tag{II-7-2}$$

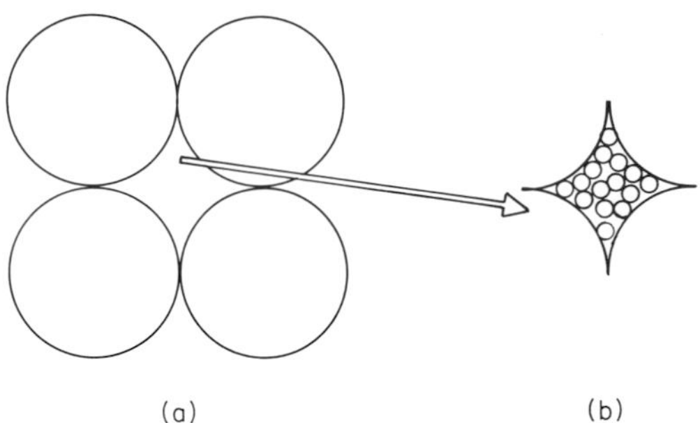

Fig. II-23 Voids of (a) coarse powder [1 cm weighs, in grams, $\rho_c' = \rho_c(1 - \epsilon_c)$] being occupied by (b) a fine powder (volume in cm^3, ϵ_c).

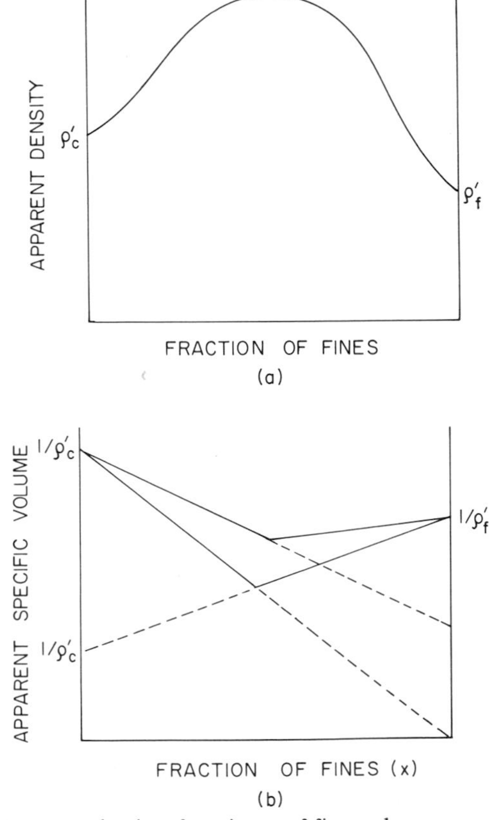

Fig. II-24 (a) Apparent density of a mixture of fine and coarse powder as a function of weight fraction x of fines. (b) Apparent specific volume curves as a function of fraction of fines. The upper curve is of the type actually experienced. The lower curves are the ones based on the simplified model of Eqs. (II-7-5) and (II-7-7).

Apparent densities can be measured as poured (cascaded) or after tapping to allow for the closest particle arrangement (tapped density ρ_t').

If fine and coarse particles are mixed and the smaller particles can *percolate*, i.e., fit in the void spaces of the larger ones, then a situation such as shown in Fig. II-23 will arise. Coarse material of 1 cm^3 weighs ρ_c' (where the subscripts and superscripts c and f denote coarse and fine) plus the amount of fines. The maximum of fines that can be introduced has a volume of ϵ_c, and hence the maximum apparent density is

$$\rho_{max}' = \rho_c' + \epsilon_c \rho_f' \quad \text{(II-7-3)}$$

Example II-7-1 A coarse powder has an apparent density of 0.9 g/cm^3 and a fine powder 0.66 g/cm^3. What is the maximum density achievable? The particle density of the coarse material is 1.35 g/cm^3.

Answer

$$\epsilon_c = 1 - 0.9/1.35 = \tfrac{1}{3}, \qquad \rho_{max}' = 0.9 + \tfrac{1}{3} \times 0.66 = 1.12 \quad g/cm^3$$

The apparent density versus fraction x of fine particles curve will have a shape as shown in Fig. II-24a. As seen from Eq. (II-7-3) the maximum should occur at

$$x_{max} = \epsilon_c \rho_f' / (\rho_c' + \epsilon_c \rho_f') \quad \text{(II-7-4)}$$

Example II-7-2 At what point will the maximum occur in the situation described in Example II-7-1?

Answer $x = 0.22/1.12 = 0.2$ g.

At $x < x_{max}$, 1 cm^3 will contain a mass (in grams) ρ_c' of coarse particles. The weight fraction of fine particles equals x, so the weight fraction of coarse matter is equal to $1 - x$, i.e., the amount of fines is $x/(1 - x)$ times that of the coarse (which is ρ_c'); hence 1 cm^3 contains

$$\rho' = \rho_c' + [x/(1 - x)]\rho_c' = [1/(1 - x)]\rho_c' \quad \text{(II-7-5)}$$

The specific volume $y = 1/\rho'$ is given by

$$y = 1/\rho' = (1/\rho_c') - (x/\rho_c') \quad \text{(II-7-6)}$$

It may be seen from this that y is linear in x with a slope of $-(1/\rho_c')$ and intercept $1/\rho_c'$. It should cut the line $x = 1$ at $y = 0$, as shown in Fig. II-24b.

Example II-7-3 Addition of fines of a particle size of 100 μm to a powder of particle size of 800 μm gives the following density data: 0%: 0.91 g/cm^3; 5%: 0.95 g/cm^3; 10%: 1 g/cm^3; 15%: 1.05 g/cm^3; 20%: 1.09 g/cm^3; 25%: 1.05 g/cm^3. What is the relation between the apparent density and the fraction of fines?

TABLE II-7

Apparent Density Data as a Function of Percent Fines

Percent fines (100x)	ρ' (g/cm^3)	y (cm^3/g)
0	0.91	1.1
5	0.95	1.05
10	1.00	1.00
15	1.05	0.95
20	1.09	0.92
25	1.05	0.95

Answer The data are tabulated in Table II-7 and the specific apparent volumes are shown (y). It is obvious that y is linear in x with the equation $y = -x + 1.1$. Note that at $x = 1$ $y = 0.1$ and not 0.

When $x > x_{max}$ the coarse powder (Fig. II-25) will simply be scattered in a bed of fine material. Here 1 g of mixture contains a mass $1 - x$ of coarse material and x of fine material. This takes up a volume of $(1 - x)/\rho_p^c$ and x/ρ_f', so

$$y = \frac{x}{\rho_f'} + \frac{1-x}{\rho_p^c} = \frac{\rho_c - \rho_f'}{\rho_f' \rho_p^c} x + \frac{1}{\rho_p^c} \tag{II-7-7}$$

Note that this line has a y intercept of $1/\rho_p^c$ and an intercept with the line $x = 1$ of $1/\rho_f'$, as shown in Fig. II-24b.

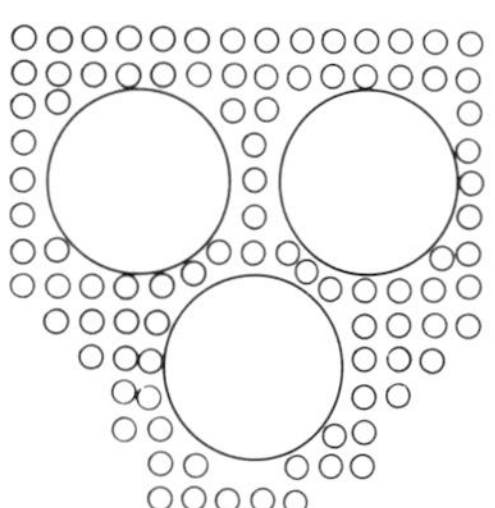

Fig. II-25 Configuration of coarse powder particles scattered in a bed of fine particles as occurs at $x > x_{max}$. [After Carstensen *et al.* (1978a).]

Example II-7-4 The particle density of the coarse material in the previous examples is 1.7 g/cm^3. At higher x values the following apparent density data are obtained: $x = 0.25$, $\rho' = 1.05$ g/cm^3; $x = 0.75$, $\rho' = 0.81$ g/cm^3; $x = 1$, $\rho' = 0.72$ g/cm^3. What is the relation between x and ρ' in this domain?

Answer The apparent specific volumes (y) are shown in Table II-8. It is obvious that the data are linear and that the equation is

$$y - 0.95 = 0.58(x - 0.25)$$

TABLE II-8

Apparent Density Data for Example II-7-4

x	ρ' (g/cm^3)	y (cm^3/g)
0.25	1.05	0.95
0.75	0.81	1.24
1.0	0.72	1.39

Note that if $x = 0, y = 0.81$; i.e., it is too large because the intercept should be $1/\rho_p^c = 1/1.7 = 0.59$ cm^3/g.

The linearity which is implied by Eqs. (II-7-6) and (II-7-7) frequently occurs, but the intercepts at the two axes of the two curves usually do not follow those predicted by the above simplified theory. Ben Aim and LeGoff (1967, 1968, pp. 1, 169) and Ben Aim *et al.* (1971) have suggested a refinement of the theory, which has been utilized by Carstensen *et al.* (1978a) to study the density behavior of mixtures of various fractions of granules (Fig. II-26). As shown in Fig. II-26 the apparent specific volume plot versus fines fraction is linear, but Eq. (II-7-4) is not obeyed for the left half of the plot because the lines do not have the required y value of zero at $x = 1$ (in fact, the lowest experimental value observed in the quoted study was 0.6). Equation (II-7-7) is also not obeyed because the y intercepts on the left-hand side of the plot in Fig. II-26 should have a value of $1/\rho_p^c$. In the case studied the particle densities of the coarse material were 1.1–1.5 g/cm^3, i.e., $1/\rho_p^c$ should have values between 0.65 and 0.9 cm^3/g, but the experimentally observed intersections were all above 1.69 cm^3/g. Carstensen *et al.* (1978a) explained this in the following fashion. Ben Aim and LeGoff (1967, 1968, pp. 1, 169) and Ben Aim *et al.* (1971) have shown that when large particles are introduced into beds of small particles, then their stacking is perturbed as shown in Fig. II-27. The layer of perturbation

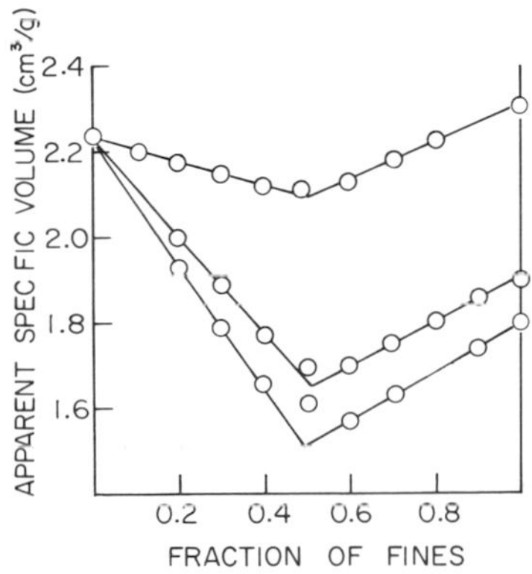

Fig. II-26 Apparent specific volume of povidone granulations. Top, a 335–715 μm mixture. Middle, a 214–715 μm mixture. Bottom, a 715–163 μm mixture. [After Carstensen *et al.* (1978a).]

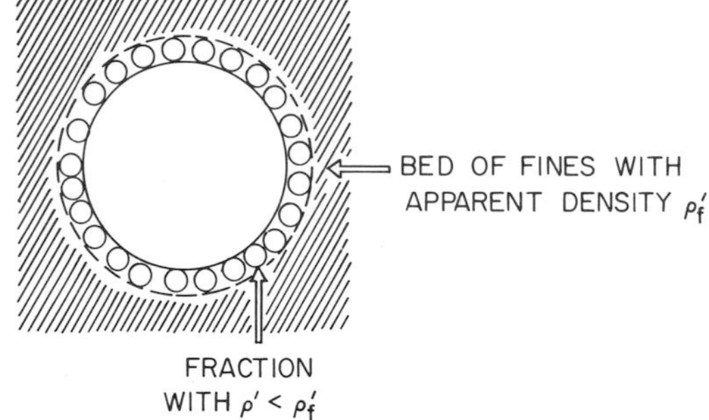

Fig. II-27 Physical situations at $x > x_{max}$. Large particles are positioned in a bed of fines, which have an apparent density of ρ_f', except for the layer which is adjacent to the large particle, where the density is smaller. [After Carstensen *et al.* (1978a).]

has an apparent density of ρ'' which differs from (and is smaller than) ρ_f'; a certain fraction a has this smaller density and a is proportional to the number of the large particles. Therefore

$$a = \alpha(1 - x) \tag{II-7-8}$$

where α is a proportionality constant. When $x > x_{max}$, 1 g of the mixture occupies a volume which (in cubic centimeters) is given by

$$y = \frac{1}{\rho'} = \frac{x - a}{\rho_f'} + \frac{a}{\rho''} + \frac{1 - x}{\rho_f^c} \tag{II-7-9}$$

Inserting Eq. (II-7-8) into Eq. (II-7-9) then gives

$$y = \frac{1}{\rho'} = \left(\frac{1}{\rho_f'} + \frac{\alpha}{\rho_f'} - \frac{\alpha}{\rho''} - \frac{1}{\rho_p^c}\right)x + \left(\frac{1}{\rho_p^c} + \frac{\alpha}{\rho''} - \frac{\alpha}{\rho_f'}\right) \tag{II-7-10}$$

The slope should therefore be smaller than that predicted by the simpler model and the intercept should be larger. This, as may be seen from Fig. II-26, is exactly what happens.

Note that the intercepts of the right line, which are denoted $y_r(0)$, are functions of the values at $x = 1$. These latter are denoted $y_r(1)$. With this nomenclature Eq. (II-7-10) can be rewritten as

$$\alpha\left[(1/\rho'') - y_r(1)\right] = y_r(0) - 1/\rho_c \tag{II-7-11}$$

This may also be written as

$$\ln(A - y_r(1)) = \ln(y_r(0) - 1/\rho_c) - \ln \alpha \tag{II-7-12}$$

Here $A = 1/\rho''$ and is used as an adjustable parameter. The best value of A is the one that gives the least sum of squares. Values of a mixture of a

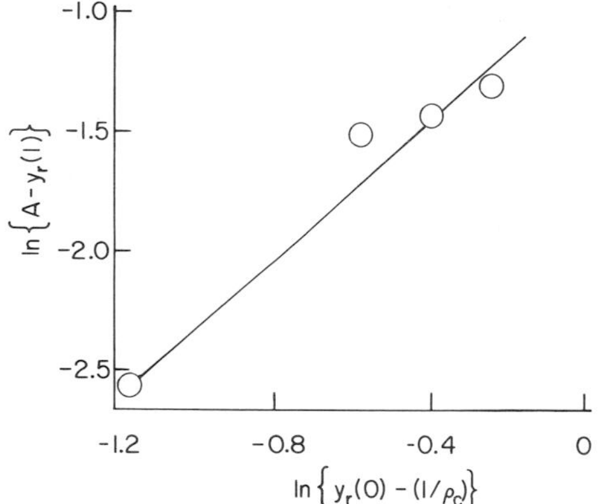

Fig. II-28 $\ln[A - y_r(1)]$ as a function of $\ln[y_r(0) - 1/\rho_c]$, with the floating parameter A having the value 2, for a povidone granulation.

coarse fraction of particles of size 750 μm and fine particles of size 163 μm are shown treated in this fashion in Fig. II-28, where it is seen that the value $A = 2$ affords the best fit to an equation of the type of Eq. (II-7-12).

Example II-7-5 In the data in the example shown in Example II-7-4, what is the value of α if $\rho'' = \frac{1}{2}\rho_f'$?

Answer $(1/\rho_p^c) + (\alpha/\rho'') - (\alpha/\rho_f') = (1/1.7) + (2\alpha - \alpha)/\rho_f' = 0.59 + (\alpha/0.66) = 0.72$, so $\alpha = 0.72 \times 0.07 = 0.06$.

When $x < x_{max}$ the cascading causes some smaller particles to lodge themselves between the larger particles. If a mass (in grams) z of fines is added to 1 cm^3 of coarse particles (which weigh ρ_c'), then this lodging will cause the volume to expand to $1 + \beta z$. Here β is a proportionality constant. The specific apparent volume is therefore given by

$$y = 1/\rho' = (1 + \beta z)/(\rho_c' + z) \qquad \text{(II-7-13)}$$

Note that $x = z/(\rho_c' + z)$. This is inserted into Eq. (II-7-13) to give

$$y = 1/\rho = (1/\rho_c') - [(1 - \beta)/\rho_c']x \qquad \text{(II-7-14)}$$

The slope-to-intercept ratio is $\beta - 1$, so that β can be calculated from this equation.

Example II-7-6 For the data given in Example II-7-3 what is the value of β?

Answer The slope-to-intercept ratio is $\beta - 1 = -1/1.1$, so $\beta = 0.09$.

II-8 STRESSES, STRAINS, AND TENSIONS IN POWDER BEDS

Most pharmaceutical operations involve the application of external forces to powder beds. This section will deal with stresses and strains caused by such situations. The fundamental notions of elasticity and flow will be discussed first.

II-8-1 Stress Definition

An external force is applied to a powder bed. With reference to Fig. II-29 (Jaeger, 1969), a small area δA in the powder bed has center O and a normal direction **P**, with the arrowed side being the positive side of the plane and the other side being the negative side. One can place a small xyz coordinate system at O as shown and the resultant force on δA at O, denoted δF, can be resolved into components along the coordinate direction (τ_{xz}, τ_{xy}, and σ_x, where **P** is in the x direction). The quantity

$$\lim(\delta F/\delta A) = \mathbf{P}_{\mathrm{OP}} \qquad \text{(II-8-1)}$$

is called the stress vector at point O. At each point O is a powder mass and in each direction **P** there exists a stress $\mathbf{P}_{\mathrm{OP}}$; i.e., the material on the positive side of the plane normal to **P** exerts a force of magnitude $\mathbf{P}_{\mathrm{OP}}\,\delta A$ on the material on the negative side (and vice versa a force of $-\mathbf{P}_{\mathrm{OP}}\,\delta A$).

We call σ_x a *normal stress*, while the τ's are called *shear stresses*. The notation τ_{xy} is frequently used and here the first subscript is the direction of the normal to δA. The second subscript is the direction in which it acts. Only one subscript is needed for the normal component (since it acts in a direction normal to δA). If σ is positive it is called tensile stress (since it attempts to pull the powder on the point on the positive side of the plane away from the material on the negative side), and if σ is negative it is

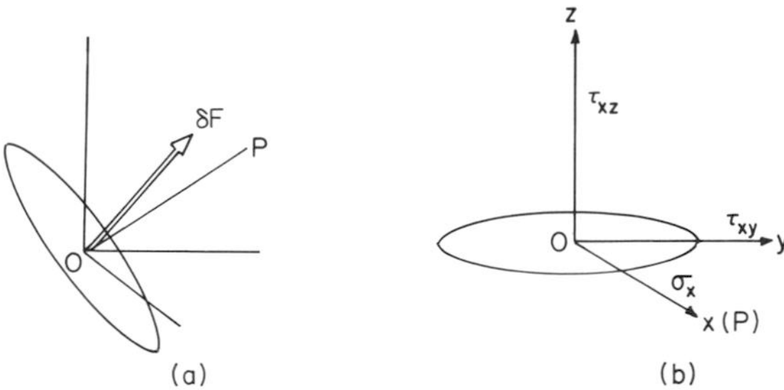

Fig. II-29 (a) Small surface element in a powder mass showing forces acting upon it, and (b) decomposition of forces into tensors in three directions.

denoted compressive stress. The stresses obtained have dimensions of Newtons per square meter. Conventionally this unit is called the Pascal (Pa), but this latter unit is reserved for traditional pressures.

One can place δA in any of three independent planes (xy, yz, zy) and hence there are nine quantities to reckon with,

$$\begin{matrix} \sigma_x & \tau_{xy} & \tau_{xz} \\ \tau_{yx} & \sigma_y & \tau_{yz} \\ \tau_{zx} & \tau_{zy} & \sigma_z \end{matrix} \qquad \text{(II-8-2)}$$

It can be shown that

$$\tau_{yx} = \tau_{xy}, \qquad \tau_{xz} = \tau_{zx}, \qquad \tau_{zy} = \tau_{yz} \qquad \text{(II-8-3)}$$

so in effect only six quantities are necessary to describe the stress at a point.

II-8-2 Principal Axes of Stress

To simplify presentation a two-dimensional situation will be considered, as shown in Fig. II-30. In this situation no quantities are dependent of z, and only the x–y plane need be considered.

In Fig. II-30 are considered the normal (σ) and shear (τ) stresses in a plane NM at an angle θ with the yz plane. With m the length of NM Eq. (II-8-4) results from projection in OP and MN directions; in the OP direction the force is σm. Expressing this in the coordinate direction and coordinate components, one sees that in the OP direction the x component contributes $\tau_{yx} \cos\theta + \sigma_y \sin\theta$ and acts on an "area" of $m \sin\theta$, so that

$$m\sigma = (\tau_{yx} \cos\theta + \sigma_y \sin\theta)m \sin\theta + (\sigma_x \cos\theta + \tau_{xy} \sin\theta)m \cos\theta \qquad \text{(II-8-4)}$$

Since $\tau_{xy} = \tau_{yx}$, this may be written as

$$\sigma = \sigma_x \cos^2\theta + 2\tau_{xy} \sin\theta \cos\theta + \sigma_y \sin^2\theta \qquad \text{(II-8-5)}$$

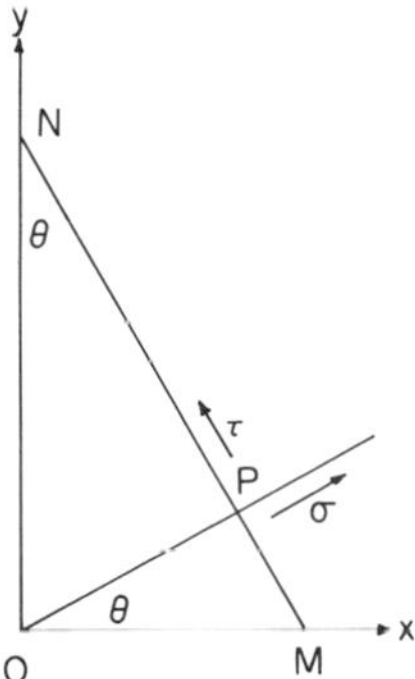

Fig. II-30 Two-dimensional example of stress components acting upon a "surface" element.

Similarly,

$$\tau = (\sigma_y - \sigma_y)\sin\theta\cos\theta + \tau_{xy}(\cos^2\theta - \sin^2\theta) \tag{II-8-6}$$

If this solution had been obtained in a different set of directions x' and y', then it can be shown that

$$\sigma_{x'} + \sigma_{y'} = \sigma_x + \sigma_y \tag{II-8-7}$$

so that the sum of σ_x and σ_y is an *invariant* with respect to rotation.

Equations (II-8-5) and (II-8-6) describe how stresses through a point change with direction. Differentiating Eq. (II-8-5) with respect to θ gives

$$\delta\sigma/\delta\theta = 2(\sigma_y - \sigma_x)\sin\theta\cos\theta + 2\tau_{xy}(\cos^2\theta - \sin^2\theta) = 2\tau \tag{II-8-8}$$

Therefore the normal stress has an extremum (and the shear stress is zero) when

$$\tan 2\theta = 2\tau_{xy}/(\sigma_x - \sigma_y) \tag{II-8-9}$$

The two directions (at right angles) defined by Eq. (II-8-9) are called *principal axes of stress*.

The following nomenclature will be used: σ_1 *and* σ_2 *are principal stresses and* $\sigma_1 > \sigma_2$. Tensile stresses are counted as positive. The impact of Eq. (II-8-9) is that when the conditions of stress at a point are known, then the direction of the principal axes can be calculated. If these are taken as axes of reference, i.e., $\sigma_x = \sigma_1$, $\sigma_y = \sigma_2$, and $\tau_{xy} = 0$, then Eqs. (II-8-5) and (II-8-6) become

$$\sigma = \sigma_1\cos^2\theta + \sigma_2\sin^2\theta = \tfrac{1}{2}(\sigma_1 + \sigma_2) + \tfrac{1}{2}(\sigma_1 - \sigma_2)\cos 2\theta \tag{II-8-10}$$

and

$$\tau = \tfrac{1}{2}(\sigma_1 - \sigma_2)\sin 2\theta \tag{II-8-11}$$

II-8-3 The Mohr Circle

Expressing σ and τ in any direction as a function of the principal stresses σ_1 and σ_2 can be done via Eqs. (II-8-10) and (II-8-11). But a graphical method, the Mohr circle, is more practical. A coordinate system (see Fig. II-31) is constructed with shear τ as ordinate and stress σ as abscissa. The points A and D on the abscissa correspond to σ_2 and σ_1. A semicircle is drawn through D and A, i.e., with the midpoint of the line segment AD, $B = ((\sigma_1 + \sigma_2)/2, 0)$, as center. To find (σ, τ) in any direction θ, the angle $DBC = 2\theta$ is drawn. The point C represents (σ, τ) in the θ direction, because the distance AC′ (in Fig. II-31) is AB + BC′ (accounting for the fact that the latter is negative, i.e., BC′ = −C′B). This equals $(\sigma_1 + \sigma_2)/2 + \overline{BC}\cos(2\theta)$, where $\overline{BC}$ is the radius and $\cos(2\theta)$ is negative

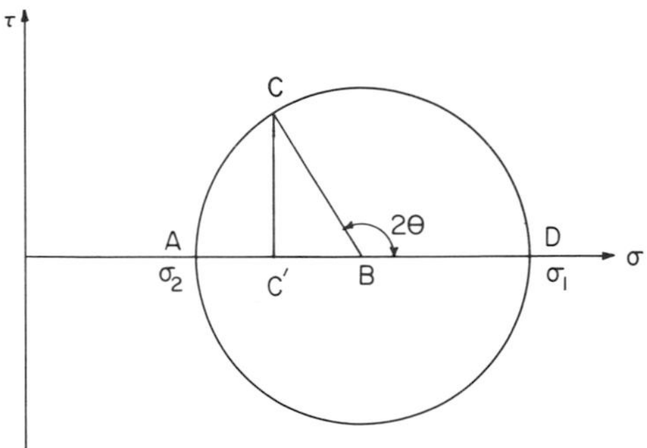

Fig. II-31 Mohr circle.

(since 2θ in the case shown is above 90°):

$$AC = [(\sigma_1 + \sigma_2)/2] + [(\sigma_1 - \sigma_2)/2] \cos 2\theta = \sigma$$

Similarly,

$$CC' = BC \sin(\pi - 2\theta) = [(\sigma_1 - \sigma_2)/2] \sin 2\theta = \tau$$

i.e., expressing the statements identically to Eqs. (II-8-10) and (II-8-11).

II-8-4 Powder Yield Loci

The notion of friction is illustrated in Fig. II-32. The force F required to move a body A over a surface B is proportional to the weight W of A:

$$F = \mu W \tag{II-8-12}$$

where μ is a frictional coefficient.

In powder work a similar situation exists when the *shear strength* of a powder is tested. The procedure is to sift powder into a ring, the so-called mask, level the powder off with a spatula, and then lift off the mask. A brass disk is then carefully placed on the powder. The desired load σ (N m^{-2}) is obtained by adding weights to the disk. Tension is applied to a thread by turning a connecting Jolly balance crank until the disk begins to move. This gives the shear strength as the required force divided by the powder contact area.

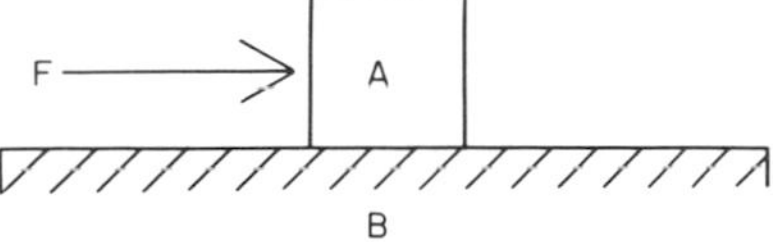

Fig. II-32 Tangential and normal force in movement of block on plane surface.

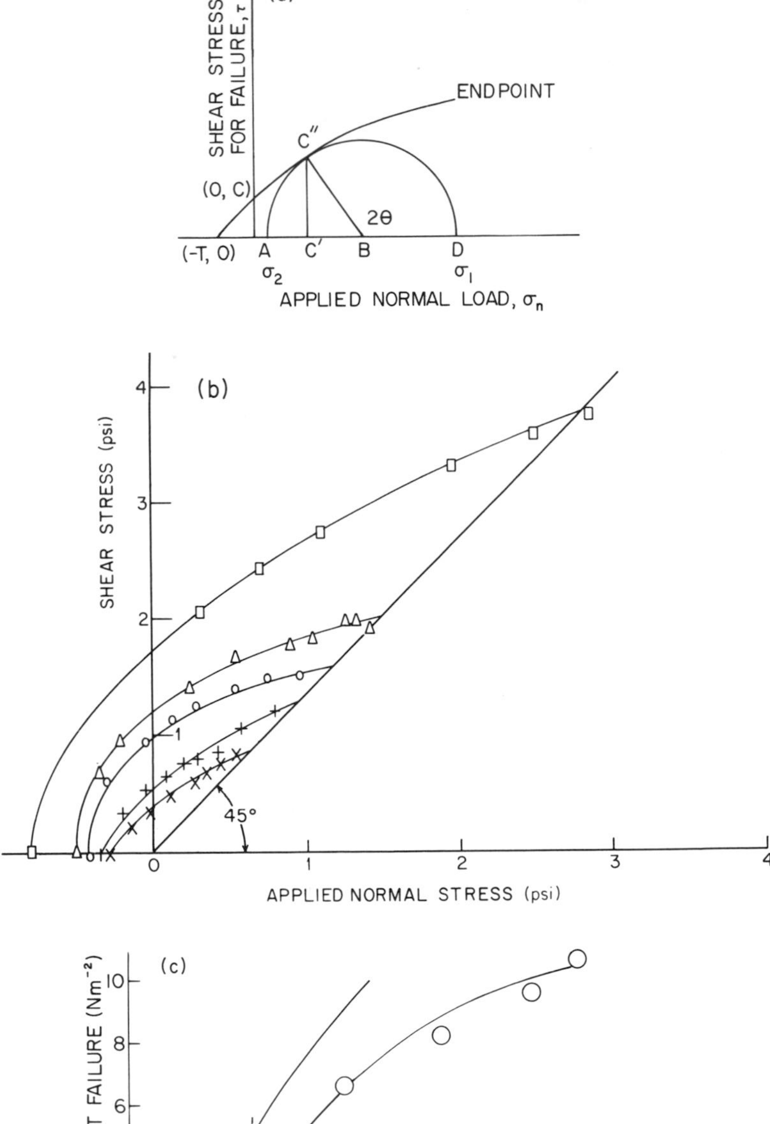

Fig. II-33 (a) Yield locus, showing the tensile strength, the cohesion, and the endpoint of the locus. (b) Yield loci for titanium dioxide. [After Williams and Birks (1967).] (c) Data from Table II-9.

If a series of normal loads σ (N m^{-2}) are applied and if the corresponding shear strengths τ (N m^{-2}) are measured, then a plot of τ versus σ will give the so-called yield locus (Fig. II-33). Various curves are obtained for various degrees of compaction (i.e., at varying apparent densities). The end point of each curve (II-33b) is frequently called the steady state point and is the point where the powder bed no longer expands at yield. Williams and Birks (1967) have shown that these points fall on a straight line and that the angle that this line forms with the x axis (or rather the $\sigma + T$ axis, as will be seen later) is a type of internal angle of friction. As seen in Fig. II-33a a Mohr circle can be drawn at each point of the yield locus so that the direction of the principal axis and the two principal stresses can be calculated. The angle is called the angle of internal friction at free surface.

Example II-8-1 Kocova and Pilpel (1973) give applied normal stress-to-tensile stress data shown in Table II-9. Draw a yield locus of the two sets.

Answer The data are plotted in Fig. II-33c.

The curves are related to the *tensile strength T* of the powders. The bulk tensile strength is obtained in an apparatus such as shown in Fig. II-34. Cylinders 1 and 2 are made of aluminum and have inside diameter 1.9 cm and are both 5.08 cm long. Cylinder 2 rests on top of cylinder 1, which sits on the base plate. Cylinder 2 has a wire handle (much like that of a can of paint) so that it can be lifted off cylinder 1 by use of a force measuring device, in this case a Jolly balance. Cylinder 3, made of brass, serves to keep cylinders 1 and 2 aligned concentrically, and the two bars keep cylinder 3 from twisting (and in this manner disturbing the powder bed inside cylinders 1 and 2). A piston with an outer knife-edge fits in cylinder 2 and serves to compress the powder. An amount of powder is weighed out and transferred to cylinder 2. The piston and necessary load weights are

TABLE II-9

Stress and Shear at Failure[a]

Tensile strength, T (n m^{-2})	σ_N (N m^{-2})	Shear at failure τ (N m^{-2})
174	329	659
—	632	803
—	930	939
—	1087	1071
226	786	1576
—	1089	1916
—	1543	2179

[a] For data reported by Kocova and Pilpel (1973).

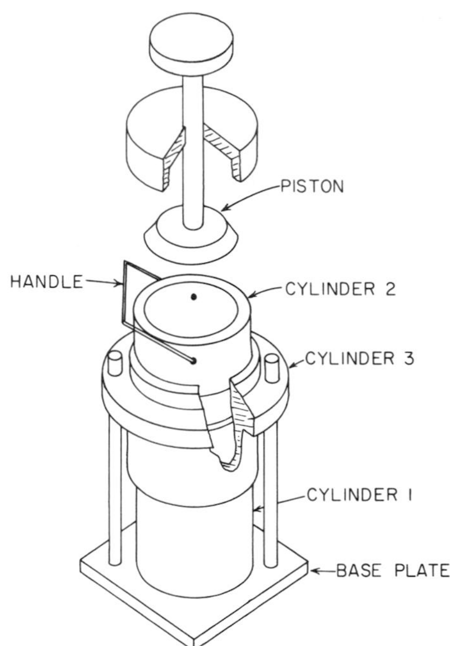

Fig. II-34 Tensile strength measuring apparatus. [After J. Nash *et al.*, *Ind. Eng. Chem. Prod. Res. Develop.* **4**, 140 (1965). Copyright (1965) by the American Chemical Society. Reprinted by permission of the copyright owner.]

inserted, the load being applied for 30–60 sec. Piston and load are then removed and the handle of cylinder 2 raised and attached to the force measuring device. The tensile strength is found by subtracting the weight of cylinder 2 and the powder that remains within it from the total force needed to lift cylinder 2 off cylinder 1 (and then dividing the resulting number by the cross-sectional area of powder). Other apparatuses (Hiestand and Wilcox, 1968) have been used for this type of measurements. Generally T is considered to be the intersection of the yield locus with the x axis. [It is the negative normal load necessary to cause failure at zero shear stress (Fig. II-33a).]

Some authors [notably Hiestand and Peot (1974)] have questioned the correctness of this yield end point. The intersection of the yield curve with the y axis (Fig. II-33a) is the *cohesional stress* C.

Example II-8-2 Draw the complete curve from Example II-8-1 and find C.

Answer It may be seen from Fig. II-33c that C is equal to 300 and 500 $\mathrm{N\,m^{-2}}$, for the two sets of data. The yield locus is a parabola of power n

(Jenike, 1961). Inclusion of C and T into the parabolic equation gives the following expression:

$$(\tau/\sigma')^n = k[(\sigma_n/\sigma') + (T/\sigma')] \qquad \text{(II-8-13)}$$

where σ' is the load at the end point of the locus (Fig. II-33b), σ/σ' is called the *reduced normal applied stress*, and τ/σ' is called the *reduced shear stress*. Another relation which would meet the criteria stated above is the so-called Warren–Springs equation, which states that

$$(\tau/C)^n = (\sigma + T)/T \qquad \text{(II-8-14)}$$

This equation may be recast as

$$(\tau/\sigma')^n = C^n(\sigma')^{1-n}(\sigma + T)/\sigma' \qquad \text{(II-8-15)}$$

so that the constant k in Eq. (II-8-13) would have to equal

$$k = C^n(\sigma')^{1-n} \qquad \text{(II-8-16)}$$

If both n, called the *shear index*, and ψ, the angle of internal friction at free surface, are constants for a powder (and they frequently are), then the powder is called *regular* (Stainforth and Berry, 1975). The shear index n is related to particle diameter d by (Farley and Valentin, 1967; Stainforth and Ashley, 1973):

$$n = 1 + 0.53d^{-2/3} \qquad \text{(II-8-17)}$$

These authors also found that the tensile strength is a function of the degree of compaction:

$$T = \phi(\rho'/\rho)^m \qquad \text{(II-8-18)}$$

where ρ' is the bulk density and ρ is the particle density. They also point out that for a huge population of powders the measured cohesional stress is about twice the tensile strength, i.e.,

$$C \sim 2T \qquad \text{(II-8-19)}$$

Note that in Example II-8-1 this is approximately true for the data of Kocova and Pilpel (1973).

Example II-8-3 For the data in the previous examples, what is the value of n?

Answer Writing the Warren–Springs equation as $\tau^n = q(\sigma + T)$ one can obtain a value of n by direct plotting. In Table II-10 are listed the logarithmic values of τ and the logarithm of $\sigma + T$. The equation for this line is $\ln \tau = 0.515 \ln(\sigma + T) + 3.267$ ($R = 0.986$). Note that $1/n = 0.515$ (the value of the slope), and it follows that $n = 1.94$.

TABLE II-10

Data from Table II-9 Treated by the Warren–Springs Equation

τ	$\ln \tau$	$\sigma_n + T$	$\ln(\sigma_n + T)$
659	6.49	503	6.22
803	6.68	806	6.69
939	6.85	1104	7.01
1071	6.98	1261	7.13

II-9 INTERPARTICULATE FORCES

Particles exert forces on one another. The forces are greater the closer they are to one another. They are usually categorized as cohesional forces.

II-9-1 Repose Angles

The effect of cohesional force is seen if a powder is poured (cascaded) onto a surface which is flat. The powder in this case will always arrange itself in a cone, a so-called heap. In the absence of a cohesive force this would not occur. This phenomenon can be explained as follows; the heap will appear as shown in Fig. II-35. Here α is the so-called angle of repose (Hiestand, 1966). As shown in the figure, a particle which is positioned on the slant of the heap is influenced by two forces: (1) the gravitational force, which is vertical and downwards, and (2) cohesional forces, which are perpendicular to the slant surface. The force of gravitation can be divided into a force which is perpendicular ($mg \cos \alpha$) and one which is tangential ($mg \sin \alpha$) to the slant surface. The particle as shown is at rest but an infinitesimal force would make it roll down the side of the slant. Recall, as shown in Fig. II-35c, that the force τ which is necessary to initiate movement of an object along the surface is proportional to the

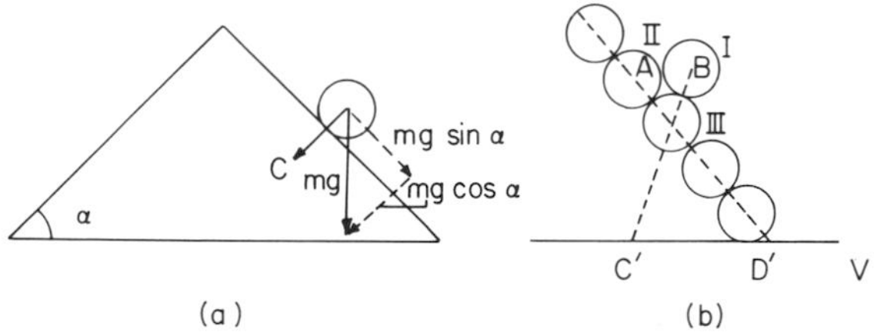

Fig. II-35 (a) Schematic of the effect of gravity and cohesive force on a particle on the slant of a heap. (b) The position of a sphere on the slope of a heap of equally sized spheres. [After Carstensen and Chan (1976b).]

force σ normal to the plane. For a slant surface the normal force, as shown in the figure, is the sum of $mg\cos\alpha$ (originating with the gravitational force) and C (the cohesional force). The force in the direction of the plane (originating from the gravitational force) is $mg\sin\alpha$. This leads to the equation

$$mg\sin\alpha = \mu(C + mg\cos\alpha) \tag{II-9-1}$$

Example II-9-1 The angle of repose of a powder is 29°, the internal friction is $\mu = 0.4$, and the powder is 70 mesh (200 μm). The density of the powder is 1.4 g/cm^3. What is the cohesive force?

Answer $\sin 29° = 0.48$. $\cos 29° = 0.87$. Mass $m = 1.4\pi(0.02^3/6) = 5.9 \times 10^{-6}$ g. $0.4(C + 5.9 \times 10^{-6} \times 980 \times 0.87) = 5.9 \times 10^{-6} \times 980 \times 0.48$. $C = 1.91 \times 10^{-3}$ dyn.

The effect of particle size on repose angles for monodisperse powders has been the subject of study of several investigators (Pilpel, 1964; Kaneniwa *et al.*, 1967; Nelson, 1955; Nogami *et al.*, 1965; Neuman, 1967) in the past. In general the repose angle α will decrease with increasing particle diameter d, and such equations as

$$\alpha = (q/d) + s \tag{II-9-2a}$$

(where q and s are parameters with values particular to the powder) have been suggested (Pilpel, 1964). Frequently the relation

$$\log\alpha = -n\log d \tag{II-9-3}$$

where n is a parameter characteristic for the particular powder holds; n here need not be unity. Pilpel (1964), however, found the data in Table II-11 for magnesium oxide powder and suggested the relation

$$\alpha = (1800/D) + 32.2 \tag{II-9-2b}$$

where the diameter D is measured in microns.

TABLE II-11

Data from Pilpel[a] and According to Eq. (II-9-2b)

Diameter (μm)	α [from Eq. (II-9-2b)]	Experimental value of α[a]	α [from Eq. (II-9-17)]
50	68.2	57	56.8
60	62.2	55	53.6
100	50.2	50	47.1
200	41.2	40	41.1
400	36.7	36	38.5

[a] From Pilpel (1964, Fig. 2).

Equations (II-9-2a) and (II-9-3) are excellent working equations and have been used with success. They suffer, however, from the inconsistency that $\alpha \to \infty$ as $d \to 0$. From a physically acceptable point of view, α should approach 90° (or less) as d approaches zero. Note from Table II-11 that Eq. (II-9-2b) holds well above 100 μm for magnesium oxide (but does not hold well at smaller particle sizes). On the other hand, Eq. (II-9-3) has been shown by Pilpel to hold well in the particle range 50–300 μm with an n value of 1.5. Equations (II-9-2a) and (II-9-3) are empirical and their connection with the physics of heaps is not apparent. An attempt will be made in the following to derive equations from the traditional concepts involved in heap formation.

It is noted that both C and μ are functions of d. It has been shown (Pilpel, 1964; Bradley, 1936; Hamaker, 1937; Jordan, 1954) that cohesive force is proportional to the diameter of the particle. Pilpel (1964) for instance quotes $C = 10^{-6}D$ dyn per particle for magnesium oxide, and in general

$$C = \zeta d \tag{II-9-4}$$

where ζ is a proportionality constant. Equation (II-9-4) implies that $C \to \infty$ when $d \to \infty$. If μ in Eq. (II-9-1) remains nonzero and finite, then Eq. (II-9-4) implies that $\alpha = 90°$ when $d \to \infty$; this, of course, is not in accordance with physical reality, and hence it is more likely that Eq. (II-9-4) has the form

$$C = \zeta_1 d/(\zeta_2 d + k) \tag{II-9-5}$$

where the ζ values and k depend on the particular substance. If $k \gg \zeta_2 d$ in the usual pharmaceutical range of d (i.e., 10–500 μm), then Eq. (II-9-5) will approximate Eq. (II-9-4). Equation (II-9-5) predicts that as $d \to \infty, C \to \zeta_1/\zeta_2$.

The dependence of μ on d is not intuitively obvious. If the concepts in Fig. II-35 are correct, then Eq. (II-9-1) may be written

$$\tan \alpha = \mu + \mu C/(mg \cos \alpha) \tag{II-9-6}$$

If as $d \to \infty$, $C \to \zeta_1/\zeta_2$, and if α approaches some finite angle, then

$$\tan \alpha \to \mu' > \mu_\infty \quad \text{as} \quad d \to \infty \tag{II-9-7}$$

As $d \to 0$, α approaches some angle of 90° or less. If one assumes the latter and notes that $C \to 0$ as $d \to 0$, then Eq. (II-9-6) becomes

$$\tan \alpha \to \mu_0 \quad \text{as} \quad d \to 0 \tag{II-9-8}$$

Since α $(= \tan^{-1} \mu_0) > \alpha_\infty$ $(= \tan^{-1} \mu_\infty)$, μ decreases with increasing particle size if the concepts in Fig. II-35 are supposed to correlate with the assumption that repose angles decrease with increasing particle size. The dependence of μ on d is the opposite of what would be expected intuitively, since the finer the particles the smoother the surface of the slant of the heap.

Note that μ approaches finite limits as d approaches either zero or infinity. A monotonically decreasing function which is bounded from both above and below is

$$\mu = (a_1 d + b_1)/(a_2 d + b_2) \tag{II-9-9}$$

where the a and b terms are characteristic for the powder in question. Equation (II-9-9) describes the simplest function meeting the physical restraints on a trial function for μ. In general it is necessary that $a_1 b_2 < a_2 b_1$ and that the a and b values be positive for the function to be monotonically decreasing, i.e.,

$$a_1 b_2 < a_2 b_1, \qquad a > 0, \quad b > 0 \tag{II-9-10}$$

The mass of a particle is related to its diameter d and the true density ρ (g/cm^3) of the solid by the relation

$$m = (\pi/6)\rho d^3 \tag{II-9-11}$$

For convenience the following terminology will be used:

$$mg = qd^3 \tag{II-9-12}$$

where

$$q = (\pi/6)\rho g \tag{II-9-13}$$

is a constant which is characteristic for a particular powder. It is of a magnitude of 500–1500 dyn/cm^3 for powders with densities of 1–3 g/cm^3.

Introducing Eqs. (II-9-5), (II-9-9), and (II-9-12) into Eq. (II-9-1) then yields

$$qd^3 \sin\alpha = \left[(a_1 d + b_1)/(a_2 d + b_2)\right]\left\{qd^3 \cos\alpha + \left[\zeta_1 d/(\zeta_2 d + k)\right]\right\} \tag{II-9-14}$$

It is possible to find values for a_1, b_1, a_2, b_2, ζ_1, ζ_2, and k which will make data conforming to Eqs. (II-9-2) and (II-9-3) conform to Eq. (II-9-14). As an example, the data from Pilpel (1964) as shown in Table II-11 fit the equation

$$1500d^3 \sin\alpha = \frac{0.7d + 0.0052}{d + 0.0018}\left(1500d^3 \cos\alpha + \frac{0.04d}{0.4d + 10}\right) \tag{II-9-15}$$

quite well. Note that $a_1 b_2 = 0.7 \times 0.0018 = 0.00126 < a_2 b_1 = 0.0052$ as required for μ to be monotonically decreasing in d.

The question as to why μ is a decreasing function in d probably has the answer that particle interlocking and stacking (Carstensen and Chan, 1976 a, b; Fukuzawa *et al.*, 1975) are the true interpretation of the μ term (rather than or in addition to friction). Interlocking is usually considered to affect the cohesional term (Pilpel, 1971), and it undoubtedly does. However, if one assumes that μ is primarily a function of particle interlocking, then μ would indeed decrease with increasing diameter, since interlocking is more pronounced at small particle sizes.

A final point in support of the views stated above is that the limiting angle at high d values frequently is about 30°. This is not surprising when it is noted that when the repose angle in Fig. II-35b (angle AD′C′) is less than 30°, then angle BC′V becomes oblique, i.e., sphere I rests in a stable configuration in the cavity formed by spheres II and III.

The above treatment shows that the traditional view of the physics of heaps is compatible with experimental data provided the coefficient of friction decreases with increasing particle diameter. By expressing the cohesion and frictional coefficients in terms of trial functions with proper boundary qualities, the basic equation derived from a force diagram of a heap will fit experimental data.

In the past two decades there have been a number of publications dealing with flow and angular properties of powders both in the form of individual investigations (Carr, 1965; Kaneniwa *et al.*, 1967; Fukuzawa *et al.*, 1975) and in the form of reviews such as those of Neuman (1967), Pilpel (1971), and Gillard *et al.* (1972). Both flow rates and repose angles of powders and granulations are frequently measured quantities in manufacturing and research in pharmaceutical and allied industries. Flow rates W are determined by monitoring the time required for a measured amount of solid to pass through an orifice, and repose angles α are obtained by forming a heap of the powder on a flat surface and measuring the angle of the cone side with the supporting surface. The interest in flow rates stems from the fact that the weight and weight variation of a tablet (or capsule) are in direct relation to the rate and uniformity of flow. Repose angles are frequently used to gauge the possible performance of a powder on compressing or filling equipment partly because there is a connection between α and W. This performance is, however, not only a function of W but also of the cohesion C and internal friction μ of the powder, and for powders there is a theoretical connection between α on the one hand and C and μ on the other (Hiestand, 1966; Carstensen, 1973, p. 185). Optimum amounts of glidants to be added to powders can be determined by repose angle determination (Nelson, 1955; Gold *et al.*, 1966a, b).

It is usual to measure C in shear cells and this is preferable since, as shown, e.g., by Pilpel (1964, 1971) the magnitude of the measured repose angles is frequently a function of the experimental technique and depends on such factors as the extent of consolidation of the powder prior to the formation of the heap. However, for materials of particle size above 100 μm, repose angles are still used for determination of C since such powders cannot be tested in shear cells due to crushing of the particles during the measurement.

A monodisperse powder with particle diameter d (cm) has a flow rate W (g/sec) which is a function of ρ (the true density, g/cm^3), g (the gravitational acceleration, cm/sec^2), the diameter D (cm) of the efflux orifice and of μ and C. These latter two, as mentioned, are functions of α. Attempts have been made in the literature to correlate W with α, ρ, and D only, and McDougall and Evans (1965) arrived at the following, often quoted equation on theoretical grounds:

$$W = (\pi/4)\left[g(1+3\gamma^2)/2\gamma\right]^{0.5}\rho D^{2.5} = j\left[(1/\gamma)+3\gamma\right]^{0.5} \qquad \text{(II-9-16)}$$

where $\gamma = \tan\alpha$ and where j consolidates variables. The first derivative of W with respect to γ is zero when

$$\partial W/\partial\gamma = 0.5j\left[(1/\gamma)+3\gamma\right]^{-0.5}\left[3-(1/\gamma^2)\right] = 0 \qquad \text{(II-9-17)}$$

i.e., when $\alpha = \tan\alpha = 1/\sqrt{3}$, or when $\alpha = 30°$. The second derivative

$$\partial^2 W/\partial\gamma^2 = -0.25j\left[(1/\gamma)+3\gamma\right]^{-1.5}\left[3-(1/\gamma^2)\right]$$

$$+\,0.5j\left[(1/\gamma)+3\alpha\right]^{-0.5}(2/\gamma^3) \qquad \text{(II-9-18)}$$

is larger than zero when $\gamma = 1/\sqrt{3}$ is inserted. This means that Eq. (II-9-16) predicts a minimum in the flow rate at a repose angle of 30°.

Carstensen and Chan (1976a, b) studied the correlation between flow rates and repose angles of *granulations*. Typical results are shown in Fig. II-36a. It is obvious that W exhibits a maximum, not a minimum at values of 29°–31°. This is in accordance with general experience (Carstensen, 1976), which dictates that repose angles of powders of 28°–42° constitute a good working range for pharmaceutical powders.

The different repose angles were obtained by separating different mesh cuts of granulations. Flow, of course, is a function of the mean diameter of the powder sample (Carr, 1960). The flow of the granulations is shown as a function of granule size in Fig. II-36b, and the flow exhibits the same dependence as reported elsewhere in literature for powders (Carr, 1960, 1965) and granulations (Kaneniwa *et al.*, 1967).

Finally, the repose angle α is plotted as a function of d (the mean mesh fraction diameter) in Fig. II-36c. Note that a maximum occurs; this is somewhat different, functionally, from what has been reported in literature for fine powders (Kaneniwa *et al.*, 1967, Pilpel, 1964) but is in accordance with some findings regarding coarse materials (Pilpel, 1971).

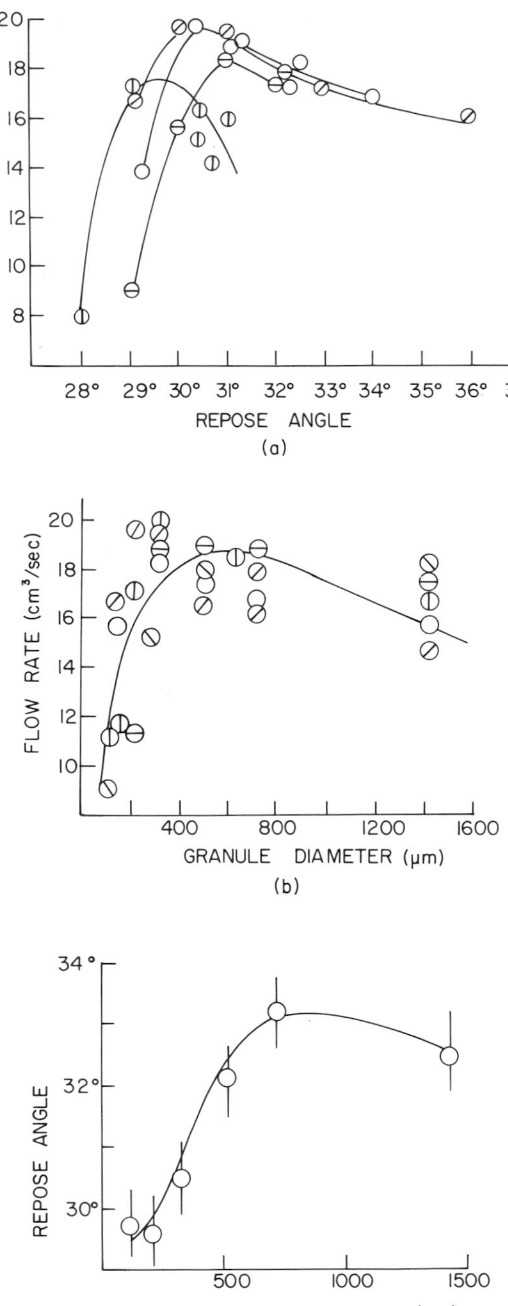

Fig. II-36 (a) Flow rate as a function of repose angle, (b) flow rate as a function of granule diameter, and (c) repose angle as a function of granule diameter, for five different granulations. [After Carstensen and Chan (1976a). Reproduced with permission of the copyright owner, the *Journal of Pharmaceutical Science*.]

For the behavior of granules, as shown in Fig. II-36, there are only two independent relations out of the three. The functional relations are

$$W = f(\alpha) \qquad \text{from Fig. II-36a} \qquad \text{(II-9-19)}$$

$$W = g(d) \qquad \text{from Fig. II-36d} \qquad \text{(II-9-20)}$$

$$\alpha = h(d) \qquad \text{from Fig. II-36c} \qquad \text{(II-9-21)}$$

That a maximum can exist in *all* of the functional relationships is best demonstrated by approximating the curves in Fig. II-36a, c by parabolas, i.e.,

$$W - W_{max} = -k_1(\alpha - \alpha_{max})^2 \qquad \text{(II-9-22)}$$

and

$$\alpha - \alpha_{max} = -k_2(d - d_{max})^2 \qquad \text{(II-9-23)}$$

where the subscript max denotes the maximum value of the dependent variable or the value of the independent variable at which the dependent variable attains its maximum. The curves in Fig. II-36a, c are *not* parabolas but functions that approximate the shape and have maxima.

If Eq. (II-9-23) is inserted in Eq. (II-9-22), then

$$W - W_{max} = -k_1 k_2^2 (d - d_{max})^4 \qquad \text{(II-9-24)}$$

which describes a curve with a maximum, i.e., the shape in Fig. II-36c.

A loosely packed population of spheres has an upper limit porosity of $\epsilon_1 = 0.785$ (Gray, 1968). As opposed to this, the porosity of a closely packed heap is more difficult to assess. Most work of this type has been done by measuring cascaded densities and tapped densities of steel balls in cylindrical containers. Scott (1962) has attributed the difficulties to wall effects, and by plotting the packing density $1 - \epsilon$ versus the reciprocal of the vessel diameter $1/D_0$ he found (1960) a linear relationship, which, when extrapolated to $1/D_0 \rightarrow 0$ (i.e., $D_0 \rightarrow \infty$) gave a porosity of 0.36. This compares favorably with the findings of Berg *et al.* (1969) that one-dimensional vertical shaking of steel spheres gives a porosity of 39% and three-dimensional vertical shaking 26% porosity, and with the findings of Newitt and Conway-Jones (1958), who found a porosity of 38% for closely packed (wet) sand.

In each of the above cases the populations of steel balls or particles have been confined by a vessel which, of course, makes it different from a heap. If a heap of spheres were cohesive, then loosely and closely packed heaps would have repose angles of 45° and about 60°, the latter being shown in Fig. II-37. Indeed, Berg *et al.* (1969) found a 60° angle of the powder remaining in a cubical container (where there is wall support). However, the 45° and 60° figures do *not* correlate with the experimental finding that the angles are in a range of 28°–36°.

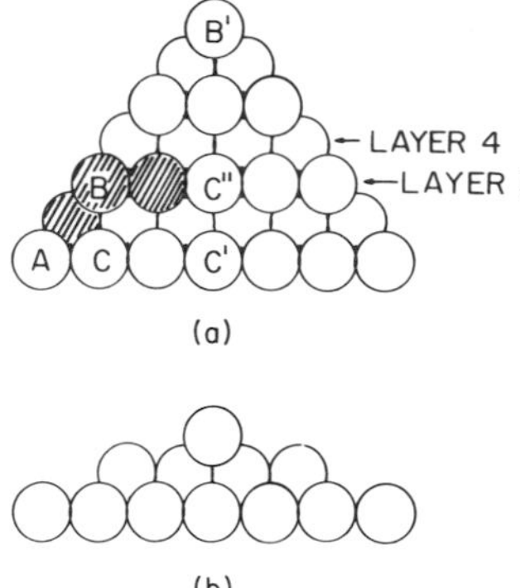

Fig. II-37 **A closely packed heap of spheres. (a) AC′B′ is unsupported by extra unloaded spheres, and (b) the left part (AC′C″) is supported by an extra unloaded sphere (e.g., A) in each layer [i.e., the cross-hatched spheres (e.g., B) disappear].**

A conical heap (Fig. II-37a) is only stable if (1) the friction between the plane support and the lower (outside) spheres is sufficiently high and (2) the friction between spheres is sufficiently high to support the outside spheres. If this is not the case, the spheres "slip" out until there are sufficient spheres in the horizontal plane without spheres over them that can serve as retainers of the first sphere in the plane that carries spheres above it. If the cross-hatched spheres are removed, the shallower cone (Fig. II-37b) results, which has two supporting spheres on each diagonal. The diagonal of the base still contains q spheres, but the first layer above contains $q - 3$ and the ith layer above the base contains $q - 3i$ spheres on the diagonal. There would be $q/3$ layers in the center and the repose angle for large values of q would equal

$$\alpha = \tan^{-1}\left[(q/3)(\sqrt{2}/2)/(q/2)\right] = \tan^{-1}(\sqrt{2}/3) = 25.2° \quad \text{(II-9-25)}$$

With these assumptions and restrictions, it can be asserted that a closely packed conical heap has a repose angle of $\alpha_c = 25.2°$ and a porosity of $\epsilon_c = 0.38$.

A closely packed heap of spheres is shown in Fig. II-37, and one (of many possible) metastable heap of loose packing as shown in Fig. II-38. The porosity of a heap, as opposed to solid populations in confining vessels, has not been covered to any extent in the literature (Gray, 1968; Carstensen, 1977).

To evaluate the porosity of a conical heap, as shown in Fig. II-37, circles were drawn as shown in Fig. II-39 with radii $1.5d, 2.5d, \ldots, qd/2$ about an arbitrarily chosen central sphere S in every second layer. For instance, in layer 3 a circle with radius $2.5d$ is drawn around the central sphere and there are $q = 5$ spheres in the diagonal layer. In alternate layers, circles of radii $d, 2d, \ldots, qd/2$ were drawn around the center of the diagonal which now is at the point P where two spheres touch. The

Fig. II-38 *Fig. II-39*

Fig. II-38 Loose packing resulting from removing every second sphere from every second layer in the supported structure shown in Fig. II-37. The depleted layers are cross-hatched. [After Carstensen and Chan (1976a). Reproduced with permission of the copyright owner, the *Journal of Pharmaceutical Science*.]

Fig. II-39 Spheres in two-dimensional close packing. In one layer, the center will be in the center of a sphere (S); in the layer above and below it, the center of the layer will be between two spheres (P). Spheres are counted as being inside the circle (part of the layer) if the center of a sphere is inside the circle. For instance, N and M are part of layer 4. The dotted sphere is not part of layer 3. [After Carstensen and Chan (1976a). Reproduced with permission of the copyright owner, the *Journal of Pharmaceutical Science*.]

number of spheres in a circle (layer) was counted with the arbitrary rule that if the center of a sphere is inside the circle, then the sphere is also "inside" the circle, i.e., part of the layer.

In Fig. II 39, N and M are part of layer 4 but not part of layer 3. In this fashion it is possible to calculate the number of spheres in the top

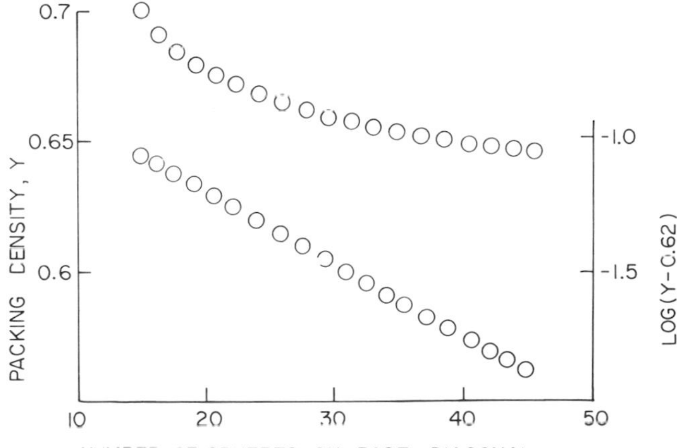

Fig. II-40 Packing densities of heaps as a function of the number of spheres on the diagonal of the base (number of layers in the heap) (upper curve). By assuming various asymptote values, subtracting these from the packing densities and plotting them semilogarithmically versus q, the best straight line, i.e., the one with the smallest residual of squares (lower curve), occurs with an asymptote value of 0.62.

layer, the next highest layer, and so on; i.e., the number of spheres can be calculated for successive values of q. The distance between each layer is $(\sqrt{2}/2)d$, so that the height of the cone is $[(q-1)(\sqrt{2}/2)+1]d$, which approaches $q\sqrt{2}/2$ for large values of q. The area of the base is $\pi(dq)^2/4$; i.e., the cone volume can be calculated. Dividing the volume of all spheres by the cone volume then gives the packing density and hence (by subtracting the latter from one) the porosity. Figure II-40 shows that the packing density approaches 0.62 (i.e., $\epsilon = 0.38$). The bed porosities found experimentally by Carstensen and Chan (1976a, b) ranged from 0.28 to 0.54.

Making reference to Fig. II-37 and noting that there are q spheres on the base diagonal, one sees that $AC' \to qd/2$ and $B'C' \to q\sqrt{2}\,d/2$ as $q \to \infty$. The repose angle hence approaches

$$\alpha = \tan^{-1}\left(qd\sqrt{2}/qd\right) = 54.7° \tag{II-9-26}$$

This value is obviously much higher than is generally encountered for cohesionless powders and granulations.

II-10 POWDER FLOW

Previous sections have dealt with powder flow of noncohesive particles. The smaller a particle becomes, the more important is cohesion, and flow is greatly affected by this. Forces of course are functions of proximity and good *glidants* [i.e., substances that improve flow or (Strickland *et al.*, 1956) size fractions which improve flow] have in common that they keep the other particles apart. Some lubricants (e.g., magnesium stearate) are generally not considered to be glidants, but are indeed good glidants. They act partly by covering the other particles with a thin film which (a) increases their distance and (b) owing to the low dielectric constant ϵ affects the magnitude of the cohesional force.

Example II-10-1 A powder with particle density 1.5 g cm^{-3} and mean diameter 50 μm has a surface area of 1 m^2/g. Its flow rate increases with addition of magnesium stearate up to an amount of 0.2%. With concentrations above this there is no further increase in flow rate. The surface area that a magnesium stearate molecule can cover is 25 Å^2 and the molecular weight of magnesium stearate is about 300. Can surface coverage be a possible explanation for this critical amount?

Answer 1 m^2 = 10,000 cm^2 can be covered by $10{,}000/(25 \times 10^{-16})$ $= 4 \times 10^{18}$ molecules $= 0.67 \times 10^{-5}$ mole = 0.002 g = 2 mg; i.e., a monolayer is formed at this percentage.

If a powder is placed in a funnel of orifice D and let flow through it, then its flow rate W can be determined. As mentioned, fine cohesive

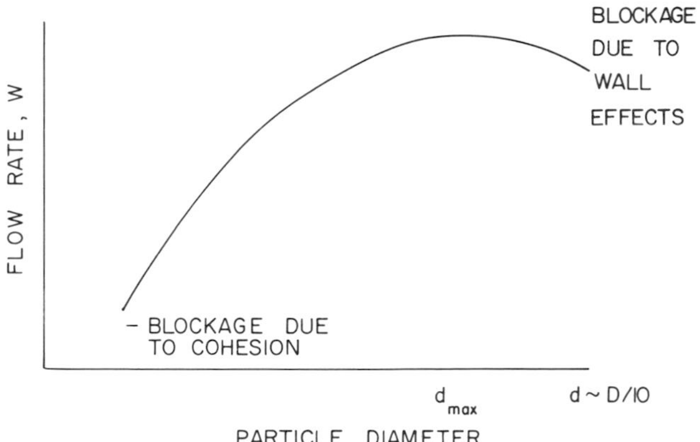

Fig. II-41 Flow rate as a function of particle diameter.

particles will flow poorly. The flow of a cohesive, monodisperse powder becomes poorer the smaller the particle diameter. Below a particular diameter d' flow will not take place at all (Fig. II-41). Making reference to Fig. II-42, let us assume that the cohesive forces act through a distance (in centimeters) of a, corresponding to n diameters (i.e., $a = nd$). If one considers one spherical particle A, it is attracted by the powder in the plane below it. There is one hemisphere that is closest to A, which is of diameter d, and hence has a surface area of $0.5\pi d^2$. The number of spherical particles (of cross section $0.25\pi d^2$) which fit onto this is $0.5\pi d^2/d^2$, since the effective area occupied by one spherical particle on the hemisphere is d^2 (akin to a square). One further diameter out the hemisphere has a surface area of $0.5\pi(2d)^2$, and hence $0.5\pi(2d)^2/d^2$ spheres fit onto it.

The force is inversely proportional to the distance squared, with proportionality constant (for convenience) denoted $2B/\pi$, so that

$$C = (B/d^2) + [B/(2d)^2]2^2 + [B/(3d)^2]3^2 + \cdots = nB/d^2 \quad \text{(II-10-1)}$$

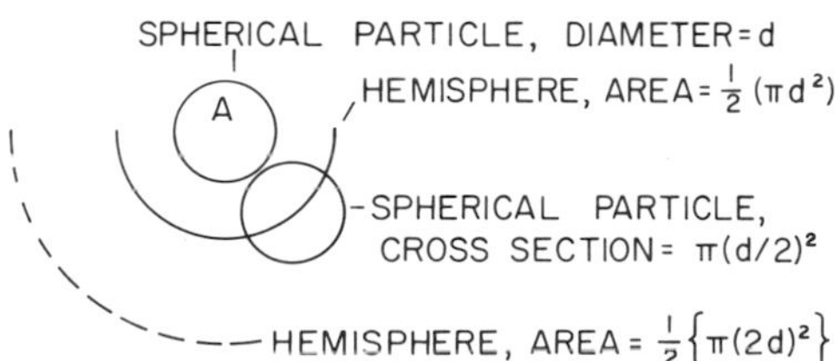

Fig. II-42 Geometry of the position of a spherical particle in a bed of spherical particles. The closest neighbors are on a sphere at a distance of (radius of) d from the center of the particle, next-nearest neighbors $2d$ from the center, etc.

Since $nd = a$, this becomes

$$C = Ba/d^3 \tag{II-10-2}$$

Hence the particle in the mass of particles (e.g., in a tube or a funnel) experiences a gravitational force $g\rho(\pi/6)d^3$ downward and a cohesional force upward, so that the total force F in the downward direction is

$$F = (\rho g\pi/6)d^3 - Ba/d^3 \tag{II-10-3}$$

and this equals zero when

$$d' = (6Ba/\rho g\pi)^{1/6} \tag{II-10-4}$$

This holds for an orifice in a funnel, but also to a degree applies to a pipe because below d' the powder mass will behave as a solid body (a "plug"). Because no (pipe or hopper) surface is smooth, the wall will support the plug at or below this diameter. Even with surfaces of great smoothness will there be a radial force. At the bottom of the tube the force will be $A\rho' lg$, where A is the cross section and ρ' is the apparent density; hence the average force on the wall is $0.5\nu A\rho' lg$, where ν is the Poisson ratio. The downward force is $A\rho' lg$, so if the wall frictional coefficient is μ then movement only occurs if

$$A\rho' lg > 0.5\nu A\rho' lg\mu$$

i.e., there will be flow only if

$$\mu < 2/\nu$$

i.e., for a high Poisson ratio, the frictional coefficient must be very small for flow to occur.

Equation (II-10-3) predicts a minimum in flow rate (at a hypothetical value d'' smaller than d'). At $d > d''$ the function is increasing, so that in general flow rate should increase with d. Once the particles are so large that cohesion no longer plays a role, the flow rate should be constant (and simply dictated by gravitation). The wall of the pipe, however, exerts a "drag" on the powder so that the "layer" closest to the efflux pipe will have a lower flow rate (W_1 particles per time unit) than the remainder [having the (higher) flow rate of W_2 particles per time unit] (Fig. II-43). The number of particles with flow rate W_2 is, per cross section

$$1 + 2\pi[(d/d) + (2d/d) + \cdots + (md/d)] = 1 + \pi m(1 + m) \tag{II-10-5}$$

It may be seen from Fig. II-43 that m is given by

$$m = (D/2d) - 1.5$$

Hence the flow rate in this region is given by

$$W_w = (p/d^2) + (q/d) + s \tag{II-10-6}$$

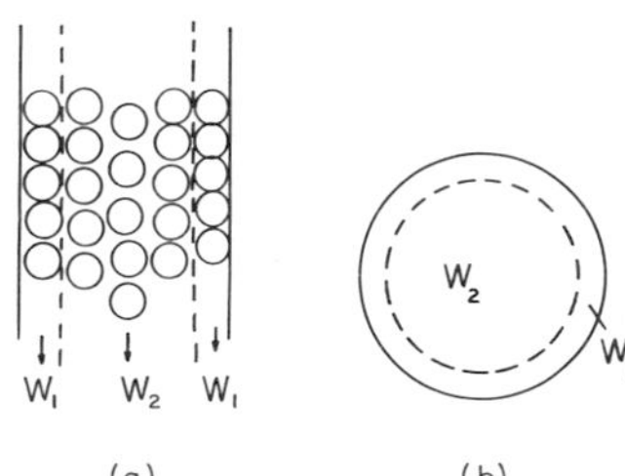

Fig. II-43 Wall effect in the flow of non-cohesive particles: (a) side view, (b) top view.

where $p = \pi W_1 D^2/4$ and $q = -\pi D(W_2 - W_1)$ and $s = W_2\{1 + (3\pi/4)\} - W_1$. This has a maximum when $d = W_2\pi D/2(W_2 - W_1)$ and this explains the final decrease in flow rate at high d values (Fig. II-41).

According to the equation of Brown and Richard, flow is also proportional to $D^{2.5}$, where D is the diameter of the efflux tube. Several phenomenological correlations have been found, e.g., by Danish and Parrott (1971) and by Jones and Pilpel (1966). These latter found that

$$W = (A^{-2.5} 15\pi) W \rho_p \sqrt{g}\, D_0^{2.5}$$

Here

$$A = 1.68d + 1.98 \qquad (d > 0.02 \text{ cm})$$
$$A = -0.85 \log d + 0.62 \qquad (0.01 < d < 0.02)$$
$$A = -0.46 \log d - 6.93 \qquad (0.003 < d < 0.01 \text{ cm})$$

Carstensen and Laughlin (1979) have extended the above principles to a dynamic situation, where the die moves below the powder bed.

REFERENCES

Ashton, M. D., and Valentin, F. H. H. (1966). *Trans. Inst. Chem. Eng. London* **44**, T166.
Ben Aim, R., and LeGoff, R. (1967). *Powder Technol.* **1**, 281.
Ben Aim, R., and LeGoff, P. (1968). *Powder Technol.* **2**, 1, 169.
Ben Aim, R., LeGoff, P., and Lelec, P. (1971). *Powder Technol.* **5**, 51.
Berg, T. G. Owe, McDonald, R. L., and Trainor, R. J., Jr. (1969). *Powder Technol.* **3**, 56, 183.
Bradley, R. S. (1936). *Trans. Faraday Soc.* **32**, 1088.
Brooke, D. (1975). *J. Pharm. Sci.* **64**, 1409.
Brunauer, S. (1961). "Solid Surfaces and the Gas-Solid Interface" (R. F. Gould, ed.), p. 15. Am. Chem. Soc., Washington, D.C.
Brunauer, S., Emmett, P. H., and Teller, E. (1938). *J. Am. Chem. Soc.* **60**, 309.
Brunauer, S., Kantro, D. L., and Weise, C. H. (1959). *Can. J. Chem.* **37**, 714.
Carr, R. (1960). *Chem. Eng. London* **67**(4), 121.
Carr, R. L. (1965). *Chem. Eng. London* **72**(2), 163.
Carstensen, J. T. (1977). "Pharmaceutics of Solids," pp. 63–85. Wiley, New York.

Carstensen, J. T. (1973). "Theory of Pharmaceutical Systems," Vol. II, Academic Press, New York.
Carstensen, J. T., and Chan, P. L. (1976a), *J. Pharm. Sci.* **65**, 1235.
Carstensen, J. T., and Chan, P. L. (1976b). *Powder Technol.* **15**, 129.
Carstensen, J. T., and Dhupar, K. (1976). *J. Pharm. Sci.* **65**, 1634.
Carstensen, J. T. Lai, T. Y. F., and Prasad, V. K. (1978b). *J. Pharm. Sci.* **67**, 1303.
Carstensen, J. T., and Laughlin, S. M. (1979). *Powder Technol.* **23**, 79.
Carstensen, J. T., and Musa, M. (1972), *J. Pharm. Sci.* **61**, 273, 1112.
Carstensen, J. T., and Patel, M. (1975). *J. Pharm. Sci.* **64**, 1770.
Carstensen, J. T., and Pothisiri, P. (1975). *J. Pharm. Sci.* **64**, 37.
Carstensen, J. T., Puisieux, F., Mehta, A., and Zoglio, M. A. (1978a). *Powder Technol.* **20**, 249.
Coppeck, J., Colvin, J., and Hume, J. (1931). *Trans. Faraday Soc.* **27**, 283.
Coppet, J., and Garner, W. (1936). *Trans. Faraday Soc.* **32**, 1739.
Dallavalle, J. M. (1948). "Micromeritics," p. 43. Pitman, New York.
Danish, F. Q., and Parrott, E. L. (1971). *J. Pharm. Sci.* **60**, 550.
Dankwerts, P. V. (1951). *Ind. Eng. Chem.* **43**, 1460.
Farley, R., and Valentin, F. H. H. (1967). *Powder Technol.* **1**, 344.
Fowler, R. (1935). *Proc. Cambridge Philos. Soc.* **31**, 260.
Fukuzawa, H., Fukuoka, E., and Kimura, S. (1975). *Yakugaku Zasshi* **95**, 859.
Garner, W., and Pike, H. (1937). *J. Chem. Soc.*, 1565.
Garner, W., and Southon, W. (1935). *J. Chem. Soc.*, 1705.
Gillard, J., Delattre, L., Jaminet, F., and Roland, M. (1972). *J. Pharm. Belg.* **27**, 713.
Gold, G., Duvall, R. N., and Palermo, B. T. (1966a). *J. Pharm. Sci.* **55**, 1133.
Gold, G., Duvall, R. N., Palermo, B. T., and Slater, J. G. (1966b). *J. Pharm. Sci.* **55**, 1291.
Goldstein, M., and Flanagan, T. (1964). *J. Chem. Educ.* **41**, 276.
Goyan, J. E. (1965). *J. Pharm. Sci.* **54**, 645.
Gray, W. A. (1968). "The packing of solid particles," p. 125. Chapman & Hall, London.
Guillory, K., and Higuchi, T. (1962). *J. Pharm. Sci.* **51**, 100.
Hamaker, H. C. (1937). *Physica* **4**, 1058.
Hatch, T., and Choate, S. P. (1929). *J. Franklin Inst.* **215**, 27.
Hiestand, E. N. (1966). *J. Pharm. Sci.* **55**, 1325.
Hiestand, E. N., and Peot, C. B. (1974). *J. Pharm. Sci.* **63**, 605.
Hiestand, E. N., and Wilcox, C. J. (1968). *J. Pharm. Sci.* **57**, 1421.
Higuchi, W. I., and Hiestand, E. N. (1963). *J. Pharm. Sci.* **52**, 67.
Hixson, A. W., and Crowell, J. H. (1931). *Ind. Eng. Chem.* **23**, 923.
Hollenbeck, R. G., Peck, G. E., and Kildsig, D. O. (1978). *J. Pharm. Sci.* **67**, 1599.
Horikoshi, I., and Himuro, I. (1966). *J. Pharm. Soc. Jpn.* **86**, 319, 324, 353, 356.
Jacobs, P. V. M., and Tompkins, F. C. (1955). "Chemistry of the Solid State" (W. E. Garner, ed.), pp. 118, 201. Butterworths, London.
Jaeger, J. C. (1969). "Elasticity, Fracture and Flow," 3rd ed., pp. 2–9. Methuen, and Science Paperbacks, London.
Jander, W. (1927). *Z. Anorg. Chem.* **163**, 1.
Jenike, A. W. (1961). Utah Eng. Exp. Station Bull. 108.
Jones, T. M., and Pilpel, N. (1966). *J. Pharm. Pharmacol.* **18**, 31, 182S, 429.
Jordan, D. W. (1954). *Br. J. Appl. Phys.* **5**, S194.
Kaneniwa, N., Ikekawa, A., and Aoki, H. (1967). *Chem. Pharm. Bull.* **15**, 1441.
Kocova, S., and Pilpel, N. (1973). *Powder Technol.* **8**, 33.
Kohlschütter, V. (1927). *Kolloid-Z.* **56**, 1169.
Lai, T. Y. F., and Carstensen, J. T. (1978). *Int. J. Pharm.* **1**, 33.
Langmuir, I. (1916). *J. Amer. Chem. Soc.* **38**, 2263.
Marshall, K., and Sixsmith, D. (1974/1975). *Drug Dev. and Ind. Pharm.* **1**, 51–71.

Marshall, K., Sixsmith, D., and Stanley-Wood, N. G. (1972). *J. Pharm. Pharmacol.* **24**, 138.
McDougall, I. R., and Evans, A. C. (1965). *Rheol. Acta* **4**, 218.
Nash, J. H., Leiter, G. G., and Johnson, A. P. (1965). *Ind. Eng. Chem. Prod. Res. Dev.* **4**, 140.
Nelson, E. (1955). *J. Am. Pharm. Assoc. Sci. Ed.* **44**, 435.
Nelson, E., Eppich, D., and Carstensen, J. T. (1974). *J. Pharm. Sci.* **63**, 755.
Neuman, B. (1967). *Adv. Pharm. Sci.* **2**, 181.
Newitt, D. M., and Conway-Jones, J. M. (1958). *Trans. Inst. Chem. Eng.* **36**, 422.
Nogami, H., Sugiwara, M., and Kimura, S. (1965). *Yakuzaigaku* **25**, 260.
Patel, M., and Carstensen, J. T. (1975). *J. Pharm. Sci.* **64**, 1651.
Pedersen, P. V., and Brown, K. F. (1975). *J. Pharm. Sci.* **64**, 1192, 1981.
Pedersen, P. V., and Brown, K. F. (1976). *J. Pharm. Sci.* **65**, 1437, 1442.
Pilpel, N. (1964). *J. Pharm. Pharmacol.* **16**. 705.
Pilpel, N. (1971). *Adv. Pharm. Sci.* **3**, 174.
Pothisiri, P., and Carstensen, J. T. (1975). *J. Pharm. Sci.* **64**, 1931.
Prescott, L. F., Steel, R. F., and Ferrier, W. R. (1970). *Clin. Pharmacol. Ther.* **11**, 496.
Prout, E., and Tompkins, F. (1944). *Trans. Faraday Soc.* **40**, 448.
Roginski, W., and Schultz, F. (1928). *Z. Phys. Chem.* **A138**, 21.
Scott, G. D. (1960). *Nature* **188**, 909.
Scott, G. D. (1962). *Nature* **194**, 956.
Short, M. P., Sharkey, P., and Rhodes, C. T. (1972). *J. Pharm. Sci.* **61**, 1733.
Stainforth, P. T., and Ashley, R. C. (1973). *Powder Technol.* **7**, 215.
Stainforth, P. T., and Berry, R. F. (1975). *Powder Technol.* **12**, 29.
Strickland, W. A., Jr., Busse, L., and Higuchi, T. (1956). *J. Am. Pharm. Assoc. Sci. Ed.* **45**, 482.
Volmer, M. (1925). *Z. Phys. Chem.* **115**, 253.
Wadell, H. (1934). *J. Franklin Inst.* **217**, 459.
Williams, J. C., and Birks, A. H. (1967). *Powder Technol.* **1**, 199.

CHAPTER

III

Two-Component Systems

Most pharmaceutical systems, of course, are multicomponent and the perturbations that are caused by such systems are in their first simplification those which are encountered in a binary system. The binary systems will be treated in the following, first from a basic point of view, and after that, from a particulate point of view.

III-1 BINARY PHASE DIAGRAMS

If a compound I and a compound II are mixed and raised to a high temperature shown by the point M in Fig. III-1 and the temperature is then let fall, then at temperature T_A crystals will start forming. These crystals are crystals of compound I and the temperature will then continue to drop with further precipitation of I until a temperature T_E is reached at point C. This is called the eutectic temperature. At this point the temperature will stay constant until all material has frozen out, and then the temperature of the now completely frozen mass can continue to drop. The cooling curve will have a shape such as shown in Fig. III-1b. The melting temperatures of I and II are T_1 and T_2, respectively, as shown in the figure, where the phases present in each area of the diagram are indicated.

Recall from the Gibbs phase rule that the number of degrees of freedom equals the number of compounds minus the number of phases

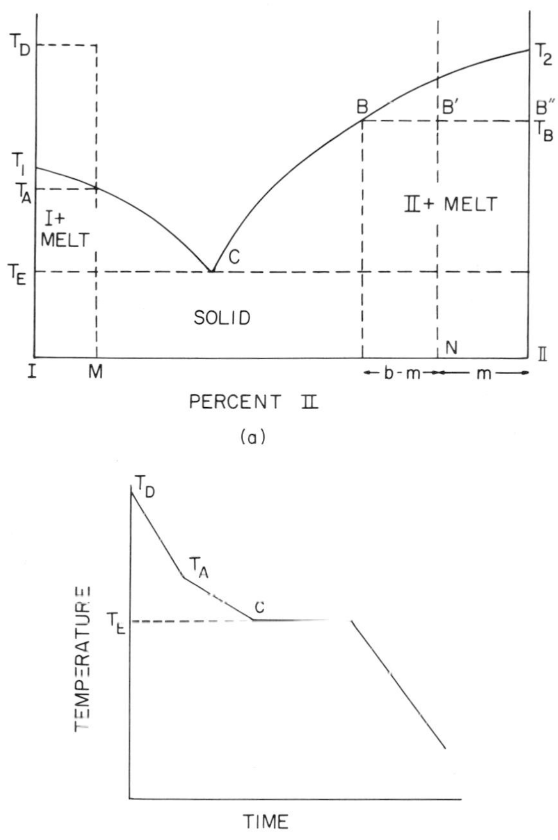

Fig. III-1 (a) Simple melting point diagram, with eutectic temperature T_E. (b) Cooling curve of composition M from temperature T_D.

plus 2, and one can see from this, for instance, that at point C there are four phases present, i.e., crystals of each type (two phases), liquid and vapor, and that therefore

$$df = 2 - 4 + 2 = 0 \qquad \text{(III-1-1)}$$

i.e., on further heat removal, further crystallization can cause no temperature change. This also shows that when the melt which started at T_D starts crystallizing out at temperature T_A only one type of crystal can precipitate out, since otherwise the number of degrees of freedom would equal zero. This crystal would have to be of the same type as crystal A, since otherwise crystallization would cause the composition of the liquid to increase, which would require an increase rather than a decrease in temperature.

Fig. III-1a shows the phases present in the various areas of the phase diagram. It may be noted that if, for instance, a solid mixture of composition N is heated (Fig. III-1a), then at temperature T_E liquid will start forming and the amount of liquid that is formed can then be calculated by the weight arm rule. For instance, at temperature T_B the amount of solid (x) and liquid ($1 - x$) present is given by

$$xm = (1 - x)(b - m) \tag{III-1-2}$$

These types of diagrams are very important in (a) lyophilization and (b) physical incompatibility. In the first case, an (unstable) solution is filled into vials, which are then frozen (i.e., brought to a temperature below the eutectic). The solid solvent (mostly water) is then evaporated off and afterwards the vials are brought back to room temperature. If not all the water is removed, then raising the temperature will cause liquid to appear at the eutectic temperature. Small amounts of liquid of course can be tolerated, but larger amounts would cause a so-called melt-back. The critical percentage of water in such a preparation can obviously be calculated from knowledge of the surface area of the solid, its solubility, and the eutectic diagram.

III-1-1 Melting Diagrams Viewed as Solubility Curves

DeLuca and Lachmann (1965) have described a theoretical basis for the simple binary melting point diagram. The mole fraction of drug is given by m, $1 - m$ is the mole fraction of water, and in this particular case

$$\ln m = (-\Delta h'/R)[(1/T) - (1/T_0)] \tag{III-1-3}$$

$$\ln(1 - m) = (-\Delta h''/R)[(1/T') - (1/T_0')] \tag{III-1-4}$$

where $\Delta h'$ is the heat of fusion of water, $\Delta h''$ is the heat of solution, and T and T' are the actual temperatures and T_0 and T_0' are the melting temperatures of the water and the drug, respectively. It should be possible to calculate the eutectic temperatures from these equations, but they assume ideality, and eutectic temperatures calculated in this fashion are not completely exact.

III-1-2 Molecular Compounds

Figure III-2 shows a type of melting point diagram which is essentially a combination of two simple melting point diagrams to the left and to the right of composition Q, which is usually a rational molecular ratio and at which particular point an actual compound exists. This compound will

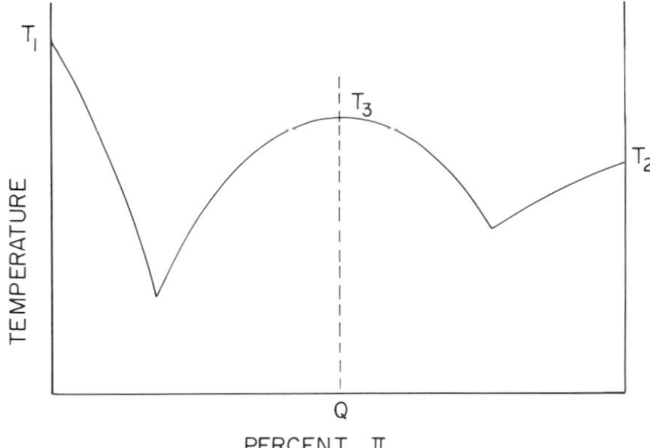

Fig. III-2 Melting point diagram with molecular compound Q.

show a clean melting point, and if, for instance, a melt at Q is cooled, then this particular compound is the one which will precipitate out. In aqueous solutions molecular compounds usually dissociate and, in this particular case, behave like weak complexes.

III-1-3 Solid Solutions

If there is a partial solubility of compound I in compound II or vice versa, then the melting point diagram will have the shape shown in Fig. III-3. In this particular case there are two types of lines: Lines AE and EC are the so-called liquidus lines, and lines AB and CD are the so-called solidus lines. The melt at 20% on line AE is in equilibrium not with solid compound I, but with a solid consisting of a small amount of compound II dissolved in compound I. The composition of the solid would be that at point F on curve AB (the solidus line).

The following situations can exist, as shown in Fig. III-4:

I. A pure compound which will simply show a sharp melting point.

II. A binary melt containing $b\%$ of compound II ($b < S_i$). This eventually yields a solid containing solid solution crystals with $b\%$ of compound II dissolved in compound I.

III. A binary melt with $\beta\%$ ($S_i < \beta < e$) of compound II. Eventually one obtains a solid which consists of solid solution crystals of composition S_i and eutectic.

IV. A eutectic composition, which will show a "normal" temperature–time curve on cooling, as if it were a pure compound. The eutecticum is a mixture of finely divided crystals of solids solutions S_i and S_j.

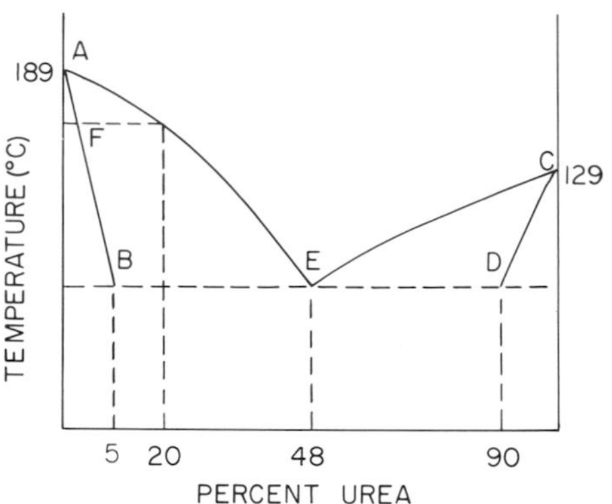

Fig. III-3 Melting point diagram where there exists partial solubility of I in solid II and of II in solid I.

V. A binary melt containing $\gamma\%$ $(e < \gamma < S_i)$ of compound II giving a solid solution of S_j and eutecticum.

VI. A binary melt with $\psi\%$ $(S_j < \psi < 100)$ of compound II giving a solid solution S_ψ on freezing.

It is appropriate to elaborate a bit on points II and IV. If it is assumed that a melt contains $100b\%$ of B, then a balance on B gives, for 100 g of melt, a mass in grams x of solid solution of a composition of S_i and a mass $100 - x$ of eutectic composition of weight fraction e. Hence

$$xS_i + (100 - x)e = 100b \qquad \text{or} \qquad x = 100(e - b)/(e - S_i) \quad \text{(III-1-5)}$$

Since x must be less than 100, it follows that S_i is smaller than b (i.e., $e - b < e - S_i$) for a eutecticum to occur in the solidified mass. Figure

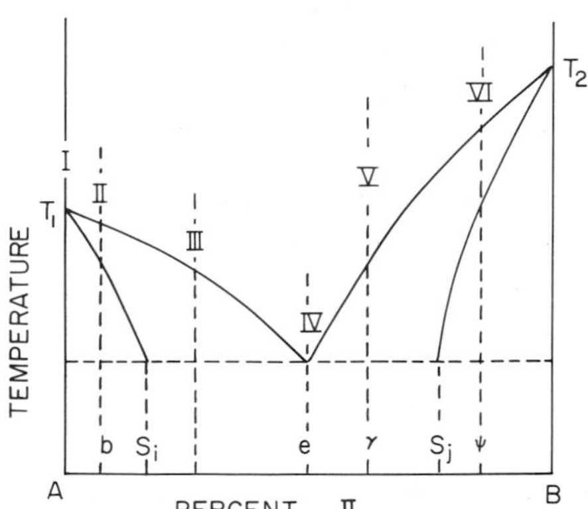

Fig. III-4 The six distinct cases of a melt cooling and solidifying in a binary system with solid solution conditions. See text for nomenclature.

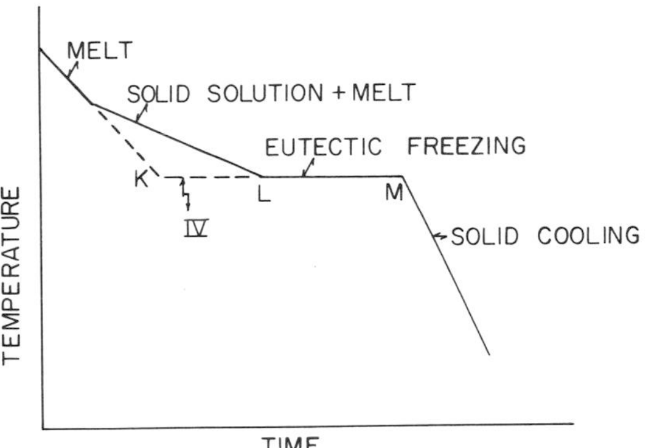

Fig. III-5 Cooling curve of the noneutectic melt and of the eutectic melt, showing the eutectic halt (KL versus KM).

III-5 shows examples of cooling curves for the situations we have denoted III and V (the sequence of four solid line segments) and for situation IV (the lower line segments, three in all, part of which are dotted lines). Of course, a pure compound will show a melting curve like that labeled IV, but the plateau temperature will be the melting point. In situations II and VI, there will be no plateau. In situations II and VI, i.e., compositions below 5% or above 90%, cooling will result only in solid solution crystals which have the same composition as the original melt. There will be no eutecticum in the solids in this situation. The above is demonstrated numerically in the case of the situation below 5%: A melt of composition m is let cool and the weight arm rule is applied, and in this situation there will be a mass (in grams) x of eutecticum (containing 48% of urea) with a weight arm of $48 - m$ and there will be $100 - x$ of solid solution crystals (with 5% urea) acting on an arm $m - 5$, so that we can write

$$(m - 5)(100 - x) = (48 - m)x \tag{III-1-6}$$

or

$$x = (m - 5)100/43 \tag{III-1-7}$$

In the case that m is smaller than 5, the conclusion is that x will be negative; i.e., no eutectic will be present. It is obviously possible to apply the same argument above 50%. Making reference to Fig. III-5 one can see that when a melt of a composition that is eutectic is let cool, then the temperature will stay constant for a certain time corresponding to line segment KM; this is called the eutectic halt and it is of longer duration than at any other composition. For instance, a composition such as that denoted III can be let cool, and in this case the amount of eutecticum which is formed is LM/KM of the amount which would have been formed from a similar amount of melt had the composition been eutectic. If one plots the eutectic halt as a function of the composition of the melt, m, then one obtains a straight line which will cut the m axis at the composition corresponding to the solid solution on the m side of the eutectic composition.

Example III-1-1 Urea and sulfathiazole form a solid solution diagram as shown in Fig. III-3. A melt containing 30 g urea and 80 g sulfathiazole is cooled slowly until it is solidified completely. What type of particles will be present and what will be the amount of the various particles in the mass of solid?

Answer The weight arm rule is applied: The right arm consists of a mass x of eutectic, acting on an arm of length $48 - 30 = 18$; the left arm consists of $100 - x$ of 5% solid solution acting on an arm that has a length of $30 - 5 = 25$.

$$18x = (100 - x)25$$

so $x = 58$ g, i.e., there are 42 g of 5% solid solution crystals and 58 g of eutecticum (finely divided mixture of 5% and 90% solid solution).

III-1-4 Consistency Relations in Solid Solution Melting Point Diagrams

There has been a certain pharmaceutical interest in solid solutions, as in the studies of Sekiguchi and Obi (1961), Sekiguchi *et al.* (1963, pp. 1108, 1123), Sekiguchi *et al.* (1964a), Goldberg *et al.* (1965), Guillory *et al.* (1969), and Chiou and Niazi (1971). The basis for solid solution systems can be obtained thermodynamically (Zhdanov, 1965, pp. 335–338) and the free energy F at absolute temperature T of a system which contains a mole fraction x of A and a fraction $1 - x$ of B can be expressed as

$$F(x, T) = K(T) + 0.5NZ\left[x^2V_{AA} + (1 - x)^2V_{BB} + 2x(1 - x)V_{AB}\right] + RT\left[x\ln x + (1 - x)\ln(1 - x)\right] \qquad \text{(III-1-8)}$$

Here N is Avogadro's number, Z the coordination number, V_{AA} the interaction energy between two A molecules, V_{BB} the interaction energy between two B molecules, V_{AB} the interaction energy between an A and a B molecule, and $K(T) = \int_0^T c\,dT - T\int_0^T (c/T)\,dT$, where c is the heat capacity. Carstensen and Anik (1976) have treated this situation. The assumptions that are made are that (a) it is only the nearest neighbor interactions that are considered, (b) it is proper to apply Stirling's formula, and (c) heat capacities are identical for both the two solids and the solid solutions

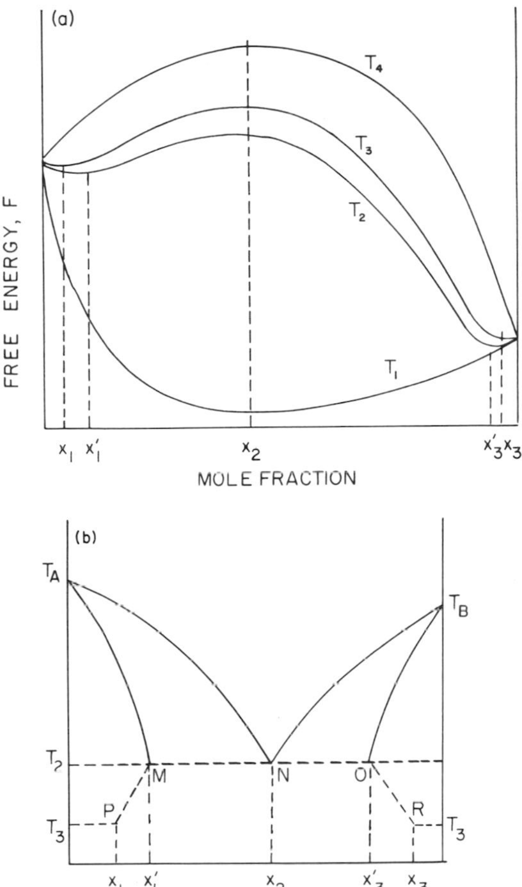

Fig. III-6 (a) Free energy versus mole fraction of a binary mixture forming a random solid solution. The indicated temperatures are labeled such that $T_1 > T_2 > T_3 > T_4$. The minima correspond to solid solution compositions at the indicated temperatures (at which the mixture is solid). In this example, T_2 could be the eutectic temperature and T_3 could be room temperature. (b) Binary melting point diagram corresponding to the energy diagram in Fig. III-6a. T_A–N and T_B–N are liquidus lines and T_A–M and T_B–O are solidus lines. [After Carstensen and Anik (1976). Reproduced with permission of the copyright owner, the *Journal of Pharmaceutical Science*.]

involved. Figure III-6 shows a plot of F as a function of x (the composition) at four different temperatures, where $T_1 > T_2 > T_3 > T_4$, where T_2 is the eutectic temperature. The assumption in Fig. III-6 is that $2V_{AB} > V_{AA} + V_{BB}$ (Ashbee, 1968). When solid solutions exist there will be three extrema (at points x_1, x_2, and x_3). There will be two minima (x_1 and x_3) at the solid solution compositions at temperature T_2 (the eutectic). It is not thermodynamically obvious that the maximum occurs at the same point as the eutectic composition. The values for x_1, x_2, and x_3 must satisfy the condition obtained from setting the first derivative of Eq. (III-1-8) equal to 0; i.e.,

$$NZ\{xV_{AA} - (1-x)V_{BB} + (1-2x)V_{AB}\} + NkT\ln[x/(1-x)] = 0 \tag{III-1-9}$$

Here k is the Boltzmann constant. If $x = x_1, x_2$, and x_3 are inserted in this, then one obtains three equations in three unknowns:

$$x_1V_{AA} + (x_1 - 1)V_{BB} + (1-2x_1)V_{AB} = -(kT/Z)\ln[x_1/(1-x_1)] \tag{III-1-10}$$

$$x_2V_{AA} + (x_2 - 1)V_{BB} + (1-2x_2)V_{AB} = -(kT/Z)\ln[x_2/(1-x_2)] \tag{III-1-11}$$

$$x_3V_{AA} + (x_3 - 1)V_{BB} + (1-2x_3)V_{AB} = -(kT/Z)\ln[x_3/(1-x_3)] \tag{III-1-12}$$

where V_{AA}, V_{BB}, and V_{AB} are the three unknowns. These three equations have no unique solution because the determinant is equal to

$$D = |x_i, (x_i - 1), (1 - 2x_i)| \tag{III-1-13}$$

which is equal to zero for all values of x_i, so that there is linear dependence among x_1, x_2, and x_3 in Eqs. (III-1-10)–(III-1-12). Taking a secular equation approach, and denoting the coefficients of dependence α_1 and α_2, it then follows that

$$\alpha_1x_1 + \alpha_2x_2 = x_3 \tag{III-1-14}$$

$$\alpha_1(x_1 - 1) + \alpha_2(x_2 - 1) = (x_3 - 1) \tag{III-1-15}$$

$$\alpha_1(1 - 2x_1) + \alpha_2(1 - 2x_2) = 1 - 2x_3 \tag{III-1-16}$$

The solutions to these equations are

$$\alpha_1 = (x_2 - x_3)/(x_2 - x_1) \tag{III-1-17}$$

$$\alpha_2 = (x_3 - x_1)/(x_2 - x_1) \tag{III-1-18}$$

For Eqs. (III-1-10)–(III-1-12) to have solutions it is necessary that the right-hand sides of the equation be subject to the same linear dependence coefficients; i.e.,

$$\alpha_1\ln[x_1/(1-x_1)] + \alpha_2\ln[x_2/(1-x_2)] = \ln[x_3/(1-x_3)] \tag{III-1-19}$$

If one now inserts Eqs. (III-1-17) and (III-1-18) and multiplies by $x_2 - x_1$, one obtains

$$(x_2 - x_3)\ln[x_1/(1-x_1)] + (x_3 - x_1)\ln[x_2/(1-x_2)] = (x_2 - x_1)\ln[x_3/(1-x_3)] \tag{III-1-20}$$

This is a consistency equation, and it should apply to the two minima and to the maximum. One knows the minima at the eutectic temperature (they occur at the solid solution compositions), so that x_2 can be found in this manner. It can then be checked whether the maximum occurs on or about the eutectic composition.

Example III-1-2 If $x_2 = 0.5$, will the curve be symmetric?

Answer Inserting $x_2 = 0.5$ into Eq. (III-1-20) gives

$$(0.5 - x_3)\ln[x_1/(1 - x_1)] + (x_3 - x_1)\ln(1.0) = (0.5 - x_1)\ln[x_3/(1 - x_3)] \quad \text{(III-1-21)}$$

The second term is 0 and the solution to the equation is therefore $x_1 = 1 - x_3$. This means that $x_2 = 0.5$ implies a symmetrical curve.

Example III-1-3 Assume that x does not occur at 0.5 and carry out similar calculations assuming that $x_1 = 0.2$ and $x_3 = 0.85$.

Answer These values are inserted in Eq. (III-1-20) and give

$$(x_2 - 0.85)\ln(0.2/0.8) + (0.85 - 0.2)\ln[x_2/(1 - x_2)] = (x_2 - 0.2)\ln[0.85/0.15] \quad \text{(III-1-22)}$$

The solution to this can be found by trial and error and is $x_2 = 0.43$.

III-2 DISSOLUTION FROM BINARY MIXTURES (MELTS)

There are frequent situations where binary mixtures occur in solid dosage forms, i.e., (a) when one uses melts and (b) when one has embedded a drug in a matrix which is insoluble in the dissolution medium. This latter case also applies when a drug is released from an ointment base which is water insoluble (e.g., petrolatum), and the former case applies when the drug releases from a base which is soluble in water (e.g., polyethylene–glycol). These latter cases have been described by Chiou and Riegelman (1971) and Chiou (1977).

In the first case, if one has a melt or a compressed mixture of two substances A and B, where both A and B are soluble in a dissolution medium (W. I. Higuchi *et al.*, 1965; Shah and Parrott, 1976), then in general steady state applies. It is assumed in the following that B dissolves more rapidly than A and that the dissolving surface recesses as shown in Fig. III-7a. For simplicity it is assumed that the total surface area is 1 cm^2 and that the material is only dissolving in a horizontal direction. The fraction of substance A present in the mixture is denoted f_A and, similarly, that of B is $f_B = 1 - f_A$. The rates with which materials A and B dissolve

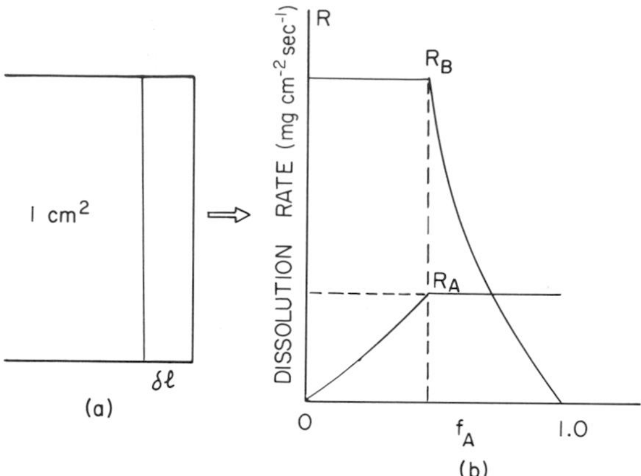

Fig. III-7 (a) Side view of a block allowing dissolution from the end surface. A 1 cm^2 surface is considered. (b) Dissolution rates R_A and R_B of two soluble substances A and B, respectively, in a molten or compressed mixture. Dissolution rates as a function of fraction of compound A.

out of 1 cm^2 of surface are denoted R_A and R_B, respectively. Since B dissolves the most rapidly at the composition in question, then in a time element δt, a depth of δl will have dissolved, so that a volume of δl will also have dissolved. Since there is no more B in the surface, i.e., since all of it is A, one can write the dissolution rate of A as

$$R_A = k_A S_A \tag{III-2-1}$$

In the time element δt there will be an amount of A dissolved ($R_A\,\delta t$) and this is f_A times the volume dissolved times the density ρ_A. The volume is dissolved is 1 δl, and so it follows that

$$R_A\,\delta t = f_A\,\delta l \rho_A, \qquad \text{i.e.,} \qquad \delta l = R_A\,\delta t / f_A \rho_A \tag{III-2-2}$$

It is possible to calculate the amount of B dissolved in a similar fashion, and then by inserting Eq. (III-2-2) one obtains

$$R_B\,\delta t = f_B\,\delta l\,\rho_B = (1 - f_A) R_A\,\delta t\,\rho_B / f_A \rho_A \tag{III-2-3}$$

i.e.,

$$R_B = (f_B/f_A)(R_A \rho_B/\rho_A) = (Q/f_A) - Q \tag{III-2-4}$$

Equations (III-2-3) and (III-2-4) imply that both of the dissolution rates are constant (steady state condition) at a particular solid composition of A/B. The equations also imply that in the case where B dissolves more rapidly than A, A's dissolution rate will not be dependent on the composition, and that the dissolution rate of B will decrease in a hyperbolic fashion [Eq. (III-2-4)] with the fraction of A.

In the case where A dissolves more rapidly than B the same line of thinking leads to the same equations with A and B inverted, i.e., in toto

$$R_A = k_A S_A \tag{III-2-5}$$

$$R_B = (Q/f_A) - Q \tag{III-2-6}$$

$$R_A = [Q'/(1 - f_A)] - Q' \tag{III-2-7}$$

$$R_B = k_B S_B \tag{III-2-8}$$

When both substances dissolve at the same rate, then one can equate Eq. (III-2-7) with Eq. (III-2-8) to give

$$f_B R_A \rho_B / f_A \rho_A = k_B S_B, \quad \text{i.e.,} \quad f_B/f_A = k_B S_B \rho_A / k_A S_A \rho_B \tag{III-2-9}$$

If one had combined Eqs. (III-2-5) and (III-2-6), one would have obtained the same set of equations as that expressed in Eq. (III-2-9).

In the case where one of the two components (A) is soluble (and hence dissolves) and where the other consitutent is an insoluble matrix, then the situation described in Fig. III-8 applies, and this situation has been treated by T. Higuchi (1963). In this case the situation exists where a matrix contains a mass A of soluble substance per cubic centimeter of preparation. If one denotes by ϵ the porosity of the matrix and by t the seconds after exposure of the matrix to the liquid, the latter will have penetrated a distance h (cm). A unit cross-sectional area is treated in the following, where we write the amount which has been released after time t (per unit area) as Q. The following equation then can be derived directly from Fig. III-8b

$$Q = Ah - 0.5\epsilon Sh \tag{III-2-10}$$

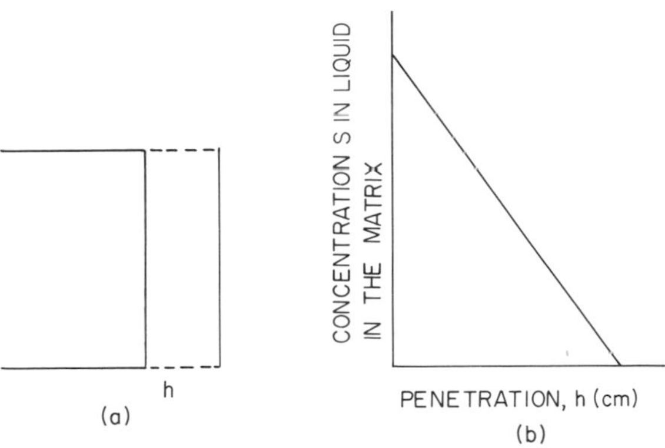

Fig. III-8 Dissolution rate of a drug A which is soluble and incorporated in an insoluble matrix (porosity = ϵ). (a) Side view of block allowing dissolution through the right surface only; h is the level of penetration at time t. (b) The concentration in the interstitial liquid as a function of distance from the exposed end of the block.

i.e.,

$$dQ/dt = (A - 0.5\epsilon S)\,dh/dt \tag{III-2-11}$$

One can write Fick's law,

$$dQ/dt = D\epsilon S/h \tag{III-2-12}$$

and combining the last two equations then yields

$$h\,dh = [D\epsilon S/(A - 0.5\epsilon S)]\,dt \tag{III-2-13}$$

This can be integrated to give

$$0.5h^2 = [D\epsilon S/(A - 0.5\epsilon S)]t \tag{III-2-14}$$

This in combination with Eq. (III-2-9) now gives

$$Q = [2D\epsilon S(A - 0.5\epsilon S)]^{1/2} t^{1/2} \tag{III-2-15}$$

This equation is the well-known Higuchi square root law. It has been repeatedly tested and verified and the works of Chien *et al.* (1974), Roseman and Higuchi (1970), and Roseman (1975) are examples of such verifications. In the latter case the work has been extended to the situation where there is diffusion of drug from the surface of the matrix placed in series with diffusion of drug through the matrix. Fessi *et al.* (1978) have extended the principles shown above to diffusion from an insoluble matrix *tablet*.

III-3 BLENDING

There is a great number of industrial blenders. An overview of these is shown in Fig. III-9. The actions are generally such that the power input will cause particles to separate from one another (overcome the cohesive forces), after which they are ready to move in relation to one another. The movement is caused by a variety of mechanisms. The ribbon blender, for instance, causes this movement by the turning of a ribbonlike blade. The drum roller does it simply by turning along its axis and depending on the movement in the surface to cause blending. Tumbling blenders, as shown, also essentially partition powder, and as will be seen later, give rise to a somewhat different than usual type of blending. Finally, the ribbon chopper blender is one which is particularly good for cohesive powders. Here a ribbon is used to let the powder pass through a small area in which high energy is input. This high energy overcomes cohesion so that agglomerates are broken up in the high-intensity area and the individual particles then blended. The principle in blending is hence that one must first allow the powders to be separated and then allow them to move in relation to one another.

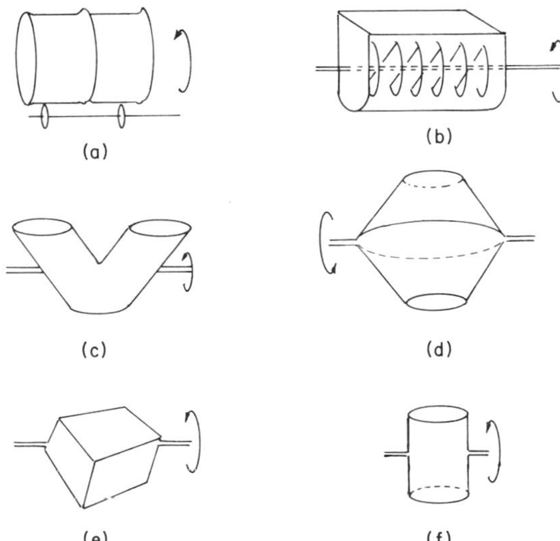

Fig. III-9 The six most common types of industrial blender: (a) horizontal barrel roller (drum roller), (b) ribbon blender (Lödige when with chopper), (c) V-blender (internal intensifier bars), (d) double cone blender, (e) rotating cube blender, and (f) end-over-end barrel roller.

Cohesive and electrostatic powders can therefore be a problem in low-power mixers because the input of energy is distributed over the mixer and hence the powder may not break up into its prime particle but may break up into agglomerates, and in this case it is only the agglomerates that are being mixed, not the individual particles. It is for this reason that the chopper ribbon blender is a good mixer for such powders. It should be pointed out also that *milling* in essence constitutes mixing because the milling operation overcomes cohesional forces between powder particles, and that therefore the milling step in some operations will add to the mixing.

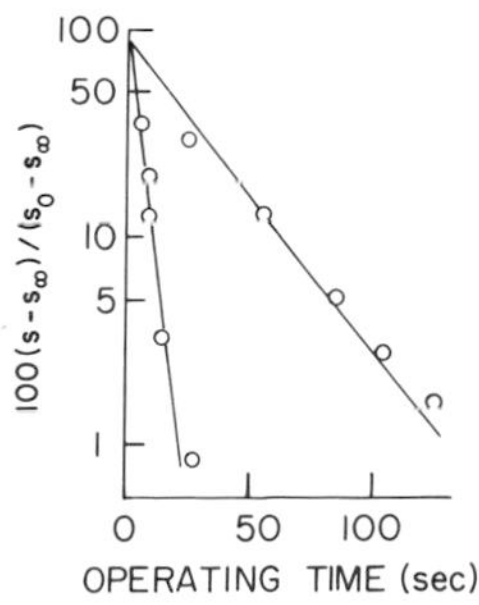

Fig. III-10 Blending of a powder in a fluid bed dryer. Top curve, air velocity of 1.0 ft/sec; bottom curve, air velocity of 1.5 ft/sec. The standard deviation is denoted by *s*. [After Nicholson and Smith (1966). Reproduced with permission of the copyright owner, *American Institute of Chemical Engineers*.]

It should finally be mentioned that fluid bed dryers (which will be covered later) give rise to very efficient mixing, but they do not overcome cohesional forces. An example (Nicholson and Smith, 1966) of this is shown in Fig. III-10. What follows deals particularly with the mathematical machinery involved in mixing.

III-3-1 Random Mixedness

It is intuitive that upon blending a fine powder will become more uniform than a coarse powder. It will be shown below that indeed the finer the powder, the higher will be the degree of uniformity that can be achieved (theoretically) by proper blending. It is worthwhile to demonstrate this by an example.

To allow handling of the math, let us assume as an extreme case that a capsule contains only three particles. Suppose the desired drug content is to be one drug particle per capsule, and that there should be two excipient particles (a fraction $1 - p$) per capsule as well. This is an exaggerated case, of course, since the capsule contains but three particles. One might now ask oneself what the probabilities are of obtaining a capsule from a random mix which contains no drug particles (Pr(0)), one particle, two particles, etc. The average drug content is $\frac{1}{3}$, so that the probability of picking three consecutive excipient particles is $\Pr(0) = (\frac{2}{3})^3 = 0.296$. The probability of picking one drug particle is similarly

$$\Pr(1) = \binom{3}{1}(\tfrac{1}{3})(\tfrac{2}{3})^2 \tag{III-3-1}$$

Here the probability of obtaining first a drug particle and then two excipient particles is given by the expression $(\frac{1}{3})(\frac{2}{3})^2$. However, there are two other ways of getting one drug particle: First pick an excipient, then a drug, then an excipient; or pick an excipient the two first times and a drug particle the third. This is the reason for the $\binom{3}{1}$ term, which is the number of ways in which one can remove one object from three if the order in which this is done is not important. This is a binomial distribution and in general is given by

$$\Pr(x) = \binom{N}{x}(p)^x(1-p)^{1-x} \tag{III-3-2}$$

where N is the number of particles in the sample, x is the number of drug particles, and p is the fraction of drug particles in the mixture. Note that the average number, in other words, the most probable number, of drug particles obtained is

$$x = pN \tag{III-3-3}$$

It can be shown that the standard deviation (in terms of particles) is given by

$$s_x = [Nx(1-x)]^{1/2} \qquad \text{(III-3-4)}$$

If two compounds are placed in a blender and mixed, then one can take samples of the mixture from time to time, and the standard deviation of the content of one component will then be a good measure of how well the two components had been mixed.

If, for the sake of simplicity, one assumes that there are 40% of component A and 60% of component B and each of these particles has a particle size of 0.42 cm, and if it is further assumed that both densities are 1.5 g/cm^3, then a 1.5 g sample would contain

$$N = (1.5/1.5)/(\pi 0.42^3/6) = 103 \quad \text{particles/sample} \qquad \text{(III-3-5)}$$

If one now were to mix the blend "completely," then one would expect that all samples would contain 40% of component A. However, as noted from Eq. (III-3-4), the best one can do is a standard deviation between samples of $(103 \times 0.4 \times 0.6)^{1/2} = 5$ particles. This essentially is a standard deviation of 5% by weight (because there are 103 particles), or based on the drug content (which is 41 particles), this would correspond to 12.5 relative %. There are therefore three types of standard deviations that one encounters in the literature:

$$s_x = [Nx(1-x)]^{1/2} \qquad \text{(III-3-4)}$$

$$s_\infty = s_x/N = [x(1-x)/N]^{1/2} \quad \text{relative standard deviation} \qquad \text{(III-3-6)}$$

$$s' = s_\infty/x = [(1-x)/Nx]^{1/2} \quad \text{coefficient of variation} \qquad \text{(III-3-7)}$$

The latter two, of course, can be expressed in percentages rather than in fractions simply by multiplying by 100.

Example III-3-1 There is 20% A and 80% B in a mixture and the particle sizes of A and B are identical. Samples of 10 g are taken, each containing 20,000 particles. What are the standard deviation, relative standard deviation, and coefficient of variation of a randomly mixed sample?

Answer

$$s_x = (40{,}000 \times 0.2 \times 0.8)^{1/2} = 80 \quad \text{particles}$$

$$s_\infty = 80/40{,}000 = 0.002 = 0.2\%$$

$$s' = (0.8/40{,}000 \times 0.2)^{1/2} = 0.01 = 1\%$$

It should be noted at this particular point that whenever N is large the binomial distribution will approximate the normal distribution and that in general we can assume that N is large, so that the assays between samples will be normally distributed. The standard deviation between the numbers that are obtained from the samples is that given by Eq. (III-3-4) [or Eq. (III-3-6) or (III-3-7)].

If two compounds A and B are placed successively in a mixer it is possible to calculate the standard deviation of a random sample taken prior to blending. Suppose that f samples are taken. A fraction x of the samples will contain all A and assay 1.0 of A. A fraction $1-x$ of the samples will contain all B (and assay 0.0 fraction of A). The mean is x fraction A, so that the relative standard deviation initially is

$$s_0 = \left[fx(1-x)^2 + f(1-x)x^2 \right]^{1/2}$$
$$= \left[fx(1-x)/(f-1) \right]^{1/2} \sim \left[x(1-x) \right]^{1/2} \qquad \text{(III-3-8)}$$

The latter is true when f is large.

During the blending operation, the relative standard deviation s will then decrease from a figure given by Eq. (III-3-8) to one given by (III-3-6), and the general form of the equation is

$$\ln Q = \ln\left[(s - s_\infty)/(s_0 - s_\infty) \right] = -kt \qquad \text{(III-3-9)}$$

where s is the standard deviation between samples at time t. An example of this is shown in Fig. III-11.

As an example, if there were 40% of A and 60% B in a mixture, then the initial standard deviation would be $(0.4 \times 0.6)^{1/2} = 0.5 = 50\%$ and the final standard deviation, if the sample size were 100 particles, would be $(0.4 \times 0.6/100)^{1/2} = 0.05 = 5\%$. A typical blending curve would then be as shown

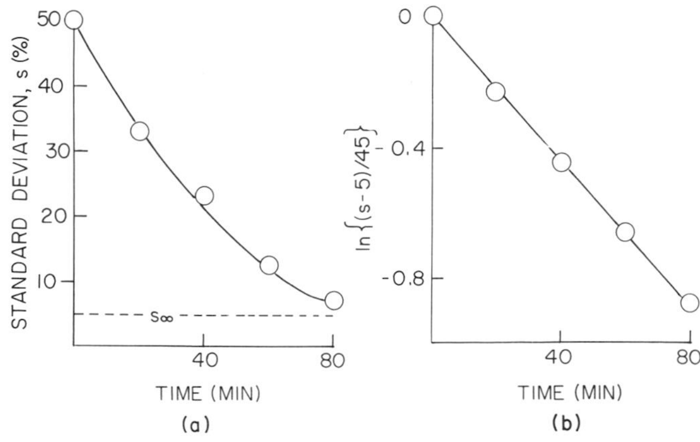

Fig. III-11 Data in Table III-1 plotted (a) as s versus time and (b) according to Eq. (III-3-9).

TABLE III-1

Blending of Powder with x = 0.4 and Sample Size N = 100

Time (min)	σ (%)	$\sigma - \sigma_\infty$	$(\sigma - \sigma_\infty)/(\sigma_0 - \sigma_\infty) = Q$	ln Q
0	50	45	1.0	0
20	33	28	0.63	− 0.46
40	23	18	0.4	−0.92
60	16.3	11.3	0.25	− 1.38
80	12.2	7.2	0.15	− 1.84

in Table III-1. It may be seen from Fig. III-11b and from the figures in the last column of Table III-1 that ln Q is linear in time with $k = 0.46/20 = 0.023\ \text{min}^{-1}$.

Blending two components directly is done when the minor component is present in at least 10% concentration. If the content is less than this (but more than 0.5%) a practice known as preblending is resorted to. When less than 0.5% concentrations apply, one frequently dissolves the minor component in a solvent, adds this to the remainder, and drives off the solvent.

When preblending is employed (and in certain other cases as well), a phenomenon known as unblending may take place. If, for simplicity, we assume that 10 parts of A are to be mixed with 90 parts of B and if the sample size is 10, then the probability of picking one A particle is

$$\Pr(1) = 10 \times 0.1 \times 0.9^9 = 0.04 = 4\%$$

If the powder is preblended 1:1 and there is a certain affinity between A and B, so that when the two parts of preblend are mixed with the remaining 8 parts of B the A + B particles move together (like one C particle), then the probability of taking an A particle is equal to the probability of taking a C particle:

$$\Pr(1) = 9 \times 0.2 \times 0.8^8 = 0.3 = 30\%$$

Since the probability of picking the "correct" composition is larger by far, the standard deviation is smaller, but if blended for long times the standard deviation will increase again and level off at the theoretical standard deviation (Fig. III-12b). Separation kinetics have been studied by Rippie and co-workers (Rippie *et al.*, 1964, 1967) and by Orr and Shotton (1973). Rippie *et al.* utilized an artificial "perfect" mix, with a standard deviation of zero, and then blended this until it eventually achieved the random state, i.e., as a function of time,

$$s' = s_\infty[1 - \exp(-k't)] \qquad \text{(III-3-10)}$$

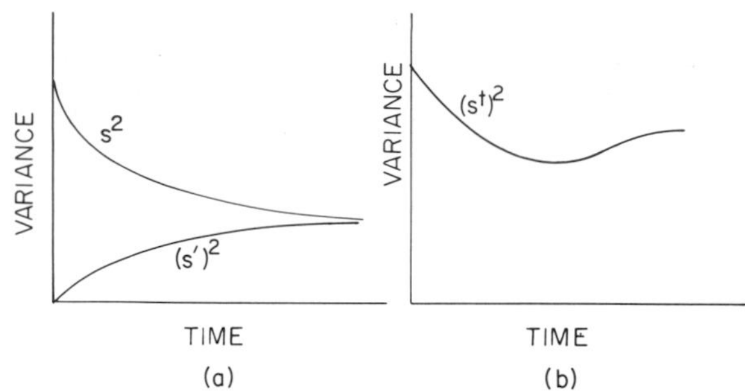

Fig. III-12 (a) Variance as a function of time for blending (s^2) and segregation (s'^2) and (b) the sum of the two.

In general there is a mixing and a segregation component in a mixing situation (Fig. III-12a) so that the total standard deviation $s^{\dagger}$ is given by

$$s^{\dagger} = s_{\infty} + (s_0 - s_{\infty})\exp(-kt) + s_{\infty} - s_{\infty}\exp(-k't) \quad \text{(III-3-11)}$$

The derivative of this with respect to time is

$$\partial s^{\dagger}/\partial t = -k(s_0 - s_{\infty})\exp(-kt) + k's_{\infty}\exp(-k't) \quad \text{(III-3-12)}$$

which can have a zero value; i.e., the curve of s versus time can have a minimum (Fig. III-12b). The argument is not rigidly correct, since (as implied in Fig. III-12) it is the variance, not the standard deviation, which is additive.

As seen in Fig. III-13b the blending is rapid when the two fractions have the same particle size, *or* when one mixing component is sufficiently fine to allow insertion into the interstices of the other component (percolation). A similar trend was found by Rippie *et al.* (1967), that the segregation rate constant is proportional to the difference in volume between the two particles. Rate constants, in contrast, are not very sensitive to differences in density.

For multicomponent mixtures there are similar expressions, which have been derived by Stange (1954, 1963), Johnson (1972), and Poole *et al.* (1964). The latter expressed the final variance of a binary mixture of polydisperse powders as

$$s_{\infty}^2 = Cy\left[y(\textstyle\sum fw)_c + C(\sum fw)_y\right]/M \quad \text{(III-3-13)}$$

where M is the mass of the sample, f is the fraction of particles of, e.g., the major component A, in a particular size range, and w is the mean weight of the particles in that size range.

Lacey (1943, 1954), Cahn *et al.* (1970), Cahn and Fuerstenau (1967, 1968), and Hogg *et al.* (1966, 1968) have considered the horizontal blend-

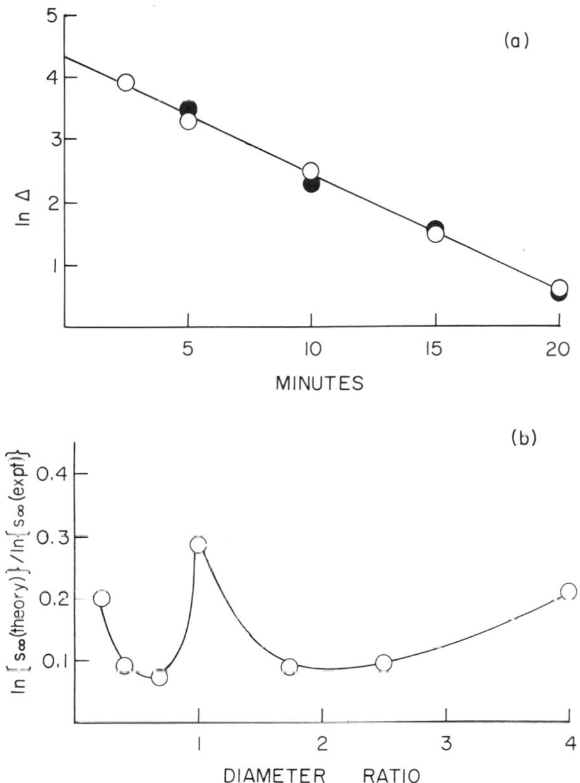

Fig. III-13 (a) Blending experiment in a V-blender. Open circles show the blending of the two components alone, and the solid circles show the blending of the two granular materials after addition of 0.5% magnesium stearate. (b) Final standard deviation (s_∞) in disized mixing as a function of the ratio of the diameters. The ordinate is the ratio of the natural logarithm of the theoretical value of s_∞ [Eq. (III-3-6)] to the logarithm of the experimental value of s_∞. [After Carstensen and Patel (1977).]

ing process as a diffusion process, and, by solving Fick's law with the appropriate initial and boundary conditions, have arrived at the equations that follow. Let D denote the diffusion coefficient (cm/sec^2), and C the fraction of one of the ingredients to be blended. The relations given are for 1:1 mixtures (the theoretical final uniform concentration $C_f = 0.5$) but can be generalized by substituting C_f for 0.5, where it occurs, and $C_f(1 - C_f)$ for 0.25. For short times the fraction C and the variance s^2 are given by

$$C = 0.5\left[1 - \operatorname{erf}\left(x/2\sqrt{Dt}\,\right)\right] \qquad \text{(III-3-14)}$$

$$s^2 = 0.25\left(1 - 4\sqrt{2Dt/\pi}\;\big/\,L\right) \qquad \text{(III-3-15)}$$

and for long times they are given by

$$C = 0.5 + (2/\pi)\exp(-\pi^2 Dt/L^2)\cos(\pi x/L) \tag{III-3-16}$$

$$s^2 = (2/\pi^2)\exp(-2\pi^2 Dt/L^2) \tag{III-3-17}$$

Experiments with actual powders of a nonspherical shape are not too abundantly reported in the literature, but where reported seem to imply that the final (random) variance is higher than given by Eqs. (III-3-6) and (III-3-13) (Kristensen, 1973; Ridgway and Segovia, 1968; Cook and Hersey, 1974). This has led Hersey (1975) and Travers (1975) to the concepts of ordered mixing and Kristensen (1973) to view mixtures by means of correlograms.

Carstensen and Patel (1977) have tested the blending of nonspherical, nonsmooth particles. Fig. III-14a shows that Eq. (III-3-14) is adhered to and Fig. III-14b shows that Eq. (III-3-17) is adhered to for nonsmooth and nonspherical particles (granules).

In other types of blenders the blending process may not be diffusion controlled. For a V-blender (Carstensen and Patel, 1977), for instance, an equation of the type of Eq. (III-3-9) [or (III-3-17), which is equivalent] is arrived at by other means than diffusion: At time zero, or prior to mixing of A and B in a 1 : 1 ratio, all the material in the left arm of the V-blender is A and all that in the right arm is B. After one rotation a fraction β of A is transferred to the right arm, so that the concentrations are $1-\beta$ and β, respectively. The difference Δ between the concentrations in the two arms is then $\Delta = 1 - 2\beta$. After a second rotation an amount of $\beta(1-\beta)$ will transfer from the left arm to the right, but an amount of β^2 will transfer from the right to the left, so that the amount in the left arm is $(1-\beta) - \beta(1-\beta) + \beta^2$, and that in the right arm will be $\beta - \beta^2 + \beta(1-\beta)$. The difference will therefore be $\Delta = (1-2\beta)^2$. Continuation of this argument shows that after N revolutions the difference between assay of the two arms should be

$$\Delta = (1 - 2\beta)^N \tag{III-3-18}$$

where concentrations are expressed in fractions. All the above arguments can be extended to ratios other than 1 : 1. The logarithmic form of Eq. (III-3-18) is

$$\ln \Delta = N \ln(1 - 2\beta) \tag{III-3-19}$$

If Δ is expressed in percent, the intercept, rather than being zero, should be $\ln 100 = 4.6$. Figure III-13a shows that Eq. (III-3-19) is adhered to well, but again, simple adherence to one equation does not guarantee that the model is correct. The variance is related to the differences between the two arms, and the data can follow both Eq. (III-3-9) and Eq. (III-3-19). It was noted,

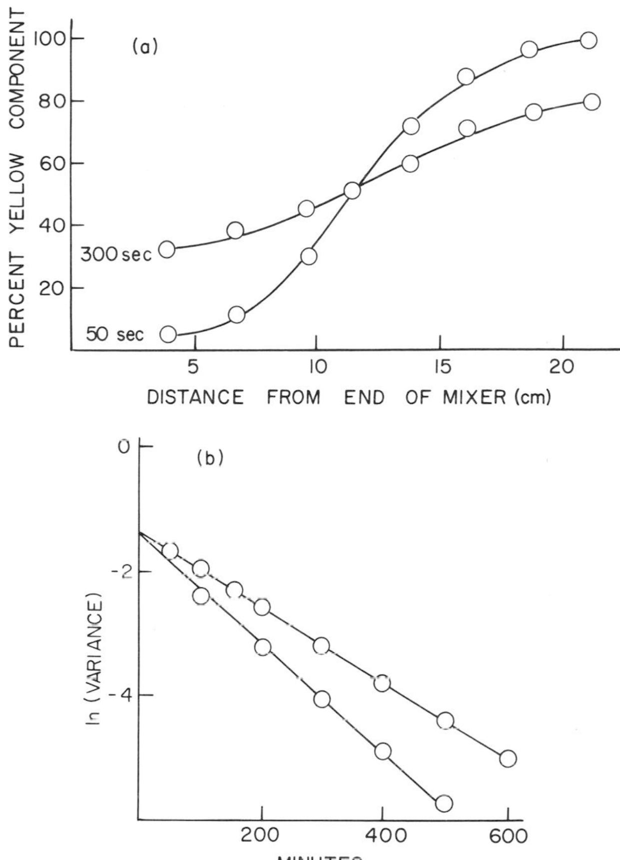

Fig. III-14 (a) Data from the blending experiment with yellow and white fractions. The blender is a horizontal, cylindrical mixer, and the content of yellow and white components is assayed at various positions (measured from the left end). The curves are cumulative normal, as predicted by Eq. (III-3-14). (b) Data from the experiment in Fig. III-14a plotted according to Eq. (III-3-17). The upper line is the blending of the two components alone, and the lower line is blending of the two granular materials after addition of 0.5% magnesium stearate. [After Carstensen and Patel (1977).]

however, that in diffusional blending the addition of lubricant increases the blending rate. As seen in Fig. III-13, this is not the case in the mixing in the V-blender. Values of β can, of course, be calculated from the slopes of such plots.

The general trends found in the blending in a horizontal cylinder are also found in V-blending. The blending rates are independent of composition, and mixing is poor and the final standard deviation high in disized mixing.

III-4 BINARY WATER SYSTEMS

Binary systems where one component is water are of a great importance in pharmaceutics, because, from a practical point of view, truly anhydrous systems of a solid character seldom exist. Many pharmaceutical parameters are strongly affected by water, such as stability, flowability, and compressibility.

III-4-1 Hydrates

The case of hydrates is one important aspect of binary water systems. Figure III-15 shows a phase diagram of Na_2HPO_4. Consider 1 mole of anhydrous salt in an evacuated container at 25°C. If sufficient moisture is let in to allow the water vapor pressure to be 6 mm Hg (or any pressure below 9.8 mm Hg), the salt will not take up any moisture; once a pressure of 9.8 mm Hg is reached, addition of moisture vapor will not result in increased vapor pressure, but moisture will be taken up by the salt: $Na_2HPO_4 + 2H_2O \rightarrow Na_2HPO_4 \cdot 2H_2O$. Once exactly 2 moles have been added beyond the point where the pressure is 9.8 mm Hg, all the salt has been converted to the dihydrate. Adding more water vapor now does not change the salt composition but raises the moisture vapor pressure until a pressure of 14.4 mm Hg has been reached. More water does not then affect the pressure but results in conversion to heptahydrate. The final equilib-

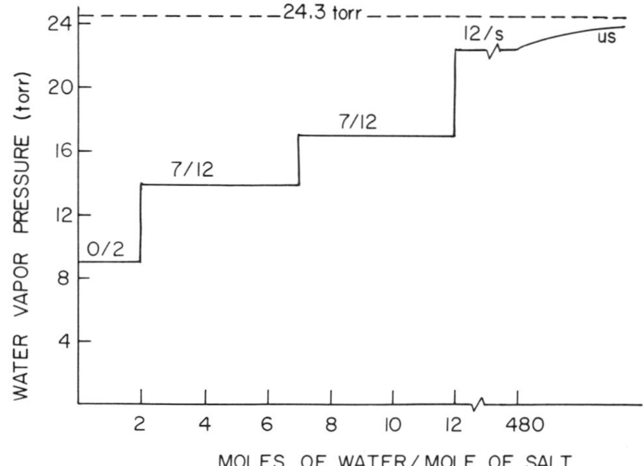

Fig. III-15 Water vapor pressures of the salt pair systems of the hydrates of Na_2HPO_4, where s denotes saturated solution, and us unsaturated solution.

rium is $Na_2HPO_4 \cdot 12H_2O + H_2O \rightarrow$ saturated solution. When all the salt is dissolved (by continuous water addition) the final curved portion (us) in Fig. III-15 will be attained.

It is important to note that a salt pair exists at a particular pressure. Therefore, in drying operations this equilibrium moisture pressure does not cast any light on the composition. If one determines an equilibrium moisture pressure of 14.4 mm Hg (50% relative humidity), there can be as little as 2 and as much as 7 mole (as little as 5% and as much as 70%) of moisture in the salt. For analytical purposes, therefore, equilibrium moisture pressure is a poor criterion. In Fig. III-15, 23 mm Hg is the vapor pressure of a saturated solution of Na_2HPO_4 and further addition of water results in liquefaction until no more crystals are left. Beyond this point, addition of water will cause (partial) condensation leading to dilution, and the vapor pressure will eventually approach that of water (24.3 mm Hg).

The equilibrium

$$\text{Salt} \cdot x\text{H}_2\text{O} + y\text{H}_2\text{O} \leftrightarrows \text{Salt} \cdot (x+y)\text{H}_2\text{O} \qquad \text{(III-4-1)}$$

has the equilibrium constant $K = P_{H_2O}^{-y}$ and a particular heat of reaction ΔH_{x+y}. The fact that P_{H_2O} is constant for the salt pair results in the horizontal line for the salt pair in Fig. III-15. The equilibrium constant K is related to temperature by a van't Hoff relation,

$$K = A \exp(\Delta H_{x+y}/RT) \qquad \text{(III-4-2)}$$

and ΔH will be different for each equilibrium (salt pair), so that the positions of the horizontal lines will change with temperature (Fig. III-16). If three hydrates exist with x, y, and z moles of water ($z > y > x$), then the

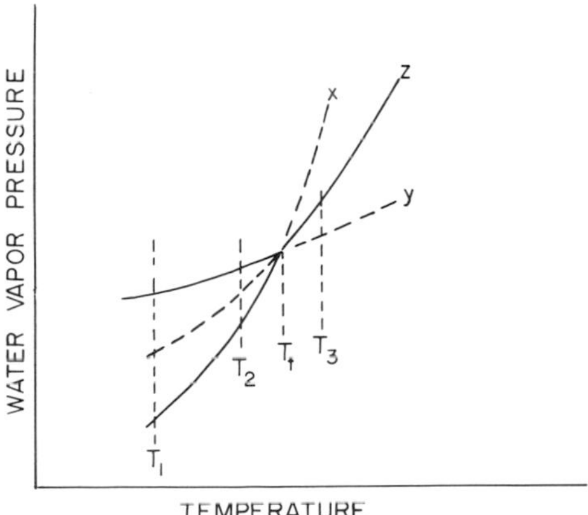

Fig. III-16 Water vapor pressure curves of three hydrates, where $x < z < y$. Above T_t only the z hydrate is stable.

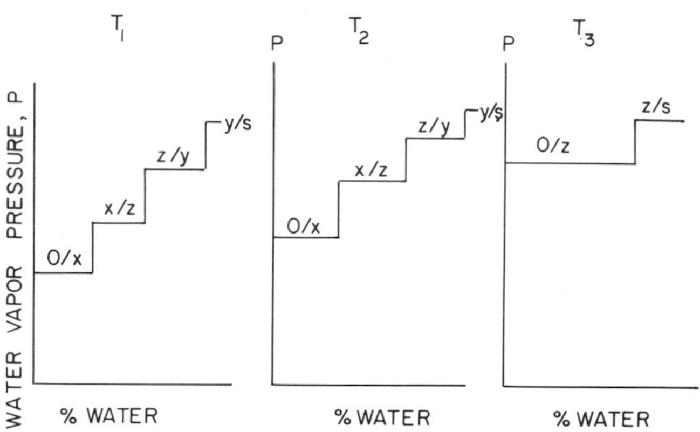

Fig. III-17 Water vapor pressure curves of the three hydrates in Fig. III-16. The nomenclature T_1, T_2, and T_3 is consistent within the two figures.

lines may come closer with increasing temperature, and at a particular temperature the lines for the *x* and *z* hydrates will disappear as shown in Fig. III-17. In reality, the lines should reverse order at the temperature T_3, but a lower hydrate cannot have a higher vapor pressure than a higher hydrate. Above the temperature T_t, only the *z* hydrate will exist; note that T_t corresponds to the temperature in Fig. III-16, where the *z* hydrate starts to be in equilibrium with saturated solution (whereas below this temperature the *y* hydrate is in equilibrium with saturated solution). Also note in Fig. III-17 the sequence by which the intermediate hydrate becomes stable at higher temperature. The opposite situation, where the *y* and *x* hydrates are stable at higher temperature and the intermediate *z* hydrate stable at lower temperature (where stability means that the phase is in equilibrium in form of a salt pair), is not thermodynamically feasible.

Rates of dehydration have only infrequently been studied in the pharmaceutical literature. There are two aspects, surface water and water held in coordinate complexes [e.g., four of the five waters of hydration in $CuSO_4 \cdot 5H_2O$; see Garner (1955)]. The former is easily removed, the latter not; for example, the last water in $MgCl_2 \cdot 6H_2O$ cannot be removed thermally (HCl is given off, leaving the oxide). When drying occurs, the surface dries first and the remainder of the drying depends on the diffusion of water through the crystal, for which reason energies of activation of drying often resemble those of diffusional processes. Three cases may occur for the residue on dehydration: (a) the crystal lattice is identical to that of the hydrate (as, e.g., in zeolites), (b) the residue has a different crystalline build up, or (c) the dehydrated material does not appear to be crystalline at all ($CuSO_4 \cdot 5H_2O$ dehydrated in vacuo).

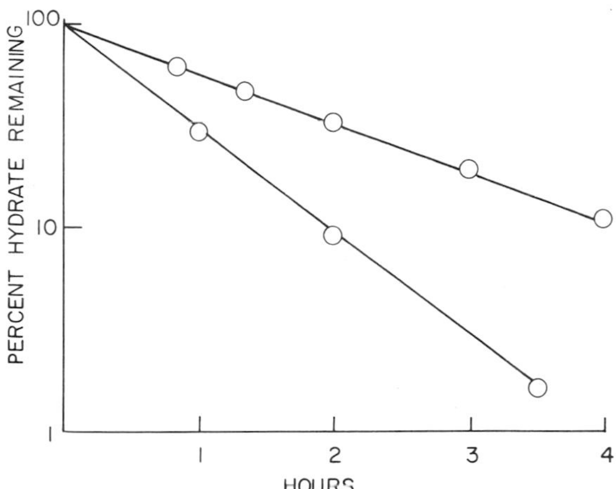

Fig. III-18 Dehydration of theophylline hydrate. The upper curve is at 40°C, and the lower curve at 50°C. [After Shefter and Kmack (1967). Reproduced with permission of the copyright owner, the *Journal of Pharmaceutical Science*.]

On "hard" drying, case (c) often occurs, which points out the fallacy of the belief that, e.g., overdried granulations can be restored to their original state by subsequent addition of water. Although the material in case (c) appears to have no crystalline structure, it mostly consists of an unclearly defined structure of extremely small crystallites (Garner, 1955).

Dehydration of theophylline was reported by Shefter and Kmack (1967) and is shown in Fig. III-18. In this case, there appears to be a straightforward first-order pattern. However, sigmoid curves often occur. Temperature is difficult to define in such experiments since the true temperature of the interface defies measurement. Thermal gravimetric analysis has given a fair amount of insight into dehydration phenomena.

Example III-4-1 Calculate the moisture contents of the dihydrate and the heptahydrate of Na_2HPO_4. The saturated solution contains 4.15 g of dodecahydrate per 100 g of water. What is the moisture content in moles of water per mole of salt?

Answer The molecular weights of Na_2HPO_4 and water are 142 and 18, respectively. The hydrates of disodium phosphate have the following water contents:

(a) anhydrous salt contains 0 g H_2O per 142 g of salt (i.e., 0%).

(2) dihydrate contains 36 g H_2O per (142 + 36) g of salt (i.e., 20.2%).

(3) heptahydrate contains 126 g H_2O per (142 + 126) g of salt (i.e., 47.0%).

(4) dodecahydrate contains 216 g H_2O per (142 + 216) g of salt (i.e., 60.3%).

The saturated solution contains 100/18 = 5.55 moles of water per 4.15 g = 0.0116 mole of dodecahydrate. The small amount of water in the latter is 0.14 mole, so a saturated solution contains (5.55 + 0.14)/0.0116 = 480 mole of water per mole of salt.

Note from Fig. III-15 that the saturated solution in the system containing solid and liquid exists in a considerable range of moisture content (60–95%). What is important is that as long as solid is present, the water vapor pressure in the vapor phase will be constant. This makes a salt solution useful, because by its use one can obtain atmospheres with a defined humidity. A list of such saturated solutions exists in common handbooks.

Relative humidity (RH) is defined as

$$\text{RH} = \frac{\text{Water-vapor pressure in the atmosphere}}{\text{Saturated water-vapor pressure}} \times 100\% \qquad \text{(III-4-3)}$$

Saturated water-vapor pressures can be found in the tables alluded to above.

Note that the equilibrium water pressures of many salt pairs do not change much with temperature, so that if one knows, for instance, the relative humidity of a salt pair at 20°C, then one can use this figure to a good approximation at temperatures to ± 10°C of the listed temperatures (in the cited case 10–30°C). This is due to the fact that the enthalpy of vaporization from a saturated solution does not differ much from that of pure water, so that if the temperature is increased, then both denominator and numerator in Eq. (III-4-3) will increase by much the same factor.

Some salt solutions have very low vapor pressures (very low relative humidities) and can actually be used to dry moist air (Kathebar units). This principle is used in certain pharmaceutical drying operations (soft-shell capsules) where LiCl solutions are employed for the drying of the moist outlet air in the drying step.

Example III-4-2 The solubility of LiCl in water is 40 g per 100 g of water and it has a molecular weight of 42.4. The vapor-pressure curve of the LiCl/H_2O system is nonideal and is given by

$$P_{H_2O} = 25.5 - 900x^2$$

Here x is the mole fraction of LiCl and P is pressure in torr. Calculate the relative humidity of the saturated solution.

Answer Since 45 g = 1.06 mole of LiCl and the 100 g of water = 5.5 mole, a saturated solution has a mole fraction of 1.06/6.60 = 0.16, so that

$P = 25.5 - 900(0.16)^2 = 2.54$. The vapor pressure of water is 25 torr, so that RH = 100(2.5/25) = 10%.

III-4-2 Equilibrium Moisture Contents

In contrast to salt pairs, certain substances (starch, montmorillonite, microcrystalline cellulose) absorb amounts of water which are continuous, smooth functions of the moisture content of the atmosphere to which they are exposed (equilibrium moisture contents, EMC). An example of this (Hollenbeck *et al.*, 1978) is shown in Fig. III-19. Note that the isotherm is smooth, is reminiscent of a BET isotherm, and shows some but not much hysteresis. Compounds of the type mentioned usually have expandable lattices allowing "insertion of water molecules between layers of host molecules (Mering, 1948). Carstensen and Su (1972) have shown how liquid adsorption isotherms of a drug (diazepam) from solution can be utilized to calculate the maximum number of moles *n* of solvent inserted in the lattice. The quantity *n* is converted to volume and this volume subtracted from the volume of the external phase. This then gives a concentration of drug dependent on *n*. Various values of *n* are tried until the drug concentrations in solution give isotherms that are inverse linear (Langmuir isotherms). The value of *n* obtained to impart linearity is comparable to those reported by Mering (1948).

Marshall and Sixsmith (1974/1975) have reported that for microcrystalline cellulose samples (Avicel 101–103) the surface areas are in the range of 10–11 $m^2 g^{-1}$ by BET, whereas Hollenbeck *et al.* (1978) found an area of 138 $m^2 g^{-1}$ for a water adsorption isotherm. The difference may be assumed due to lack of insertion of N_2 and (as mentioned) insertion of water into the host lattice.

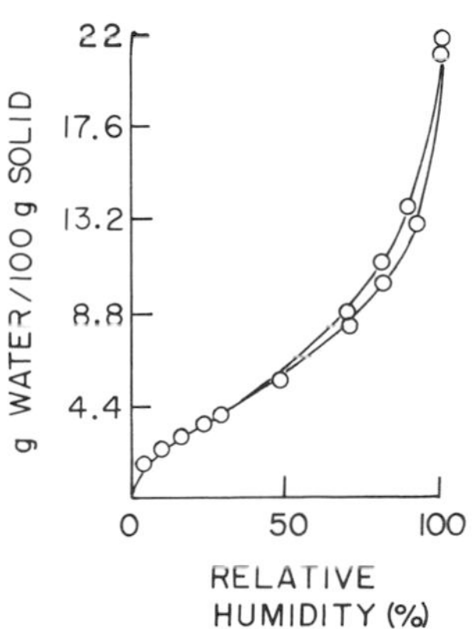

Fig. III-19 Moisture adsorption isotherm for water on microcrystalline cellulose at 25°C. [After Hollenbeck *et al.*, 1978. Reproduced with permission of the copyright owner, the *Journal of Pharmaceutical Science.*]

III-4-3 Particle Porosity

In contrast to the insertion model is the situation where fine pore volume affects water ad/absorption. Pore size distribution, as mentioned in Section II-4, can be deduced from hysteresis loops in adsorption isotherms. This is based on the fact that the vapor pressure P of a liquid with zero contact angle (i.e., $\cos\theta = 1$) condensed in a capillary with radius r is given by the Kelvin equation

$$\ln(P/P_0) = -2\gamma V/rRT \qquad \text{(III-4-4)}$$

where P_0 is the bulk vapor pressure of the adsorbate, γ is the interfacial tension (γ_{SL}) of condensed liquid (i.e., adsorbate) to solid (i.e., adsorbent), and V is the molar volume of the condensed adsorbate. During the adsorption process of, e.g., a nitrogen or water isotherm the pores will fill during adsorption, but during desorption at a pressure P' only pores of a radius above a critical radius r' will permit condensed adsorbate to escape. This radius r' is of course given by

$$r' > -2\gamma V/[RT\ln(P'/P_0)] \qquad \text{(III-4-5)}$$

The smaller P', the smaller r'.

Mercury porosimetry is reverse in the sense that high pressures are needed to force liquid mercury into a small pore since the contact angle is above 90° and its cosine is negative. The volume of mercury plus solid can be measured (in a so-called mercury porosimeter) at various pressures. The data by Marshall and Sixsmith (1974/1975) are shown in Fig. III-20. A surface area can be calculated from the pore distribution by integrating the penetration volume versus intrusion pressure graphically. This gives values for microcrystalline cellulose of the order of 1.0 $m^2\,g^{-1}$, and hence pore distribution is not the explanation for the water isotherm. However, for certain water-insoluble substances, Eq. (III-4-4) can give rise to isotherms of the conventional shape for EMC plots. Nakai *et al.* (1977, 1978a, b)

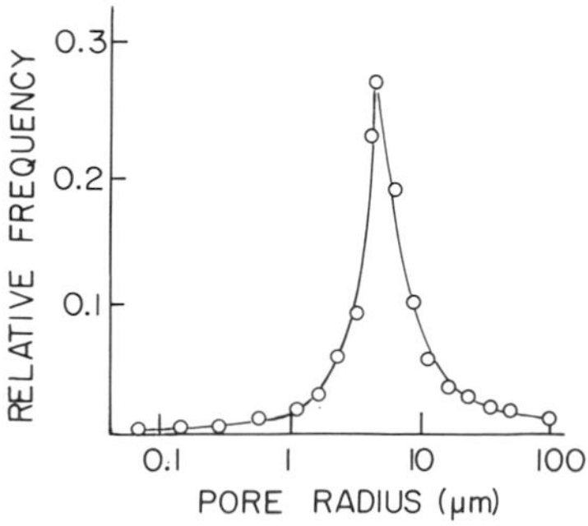

Fig. III-20 Pore size distribution of microcrystalline cellulose (Avicel PH 105). Data obtained by mercury porosimetry, using a surface tension figure for mercury of 0.47 $N\,m^{-1}$ and a contact angle of 130°. [After Marshall and Sixsmith (1974/1975). By courtesy of Marcel Dekker, Inc., New York.]

have shown that microcrystalline cellulose when mixed with a drug and then ground (milled) will substantially change the properties of the latter: up to a certain ratio, for instance, benzoic acid will not show its melting peak at differential scanning calorimetry, will dissolve more rapidly, and will sublime at a different rate. The ratio is the function of the size of the drug molecule and of the pore size distribution, i.e., the volume of pores "larger" than the molecule such that they volumetrically can accommodate the drug entity.

III-4-4 Hygroscopicity

It is a well-known fact that a term such as solubility has a definitive definition so that substances can be slightly soluble, very slightly soluble, etc. Such a definition does not exist for hygroscopicity; i.e., one cannot define a substance as being slightly hygroscopic, very hygroscopic, and so on. This is because there is both a kinetic and a thermodynamic component in the term. The considerations to follow apply to substances that show some degree of solubility. The derivations are based on a soluble substance that has no hydrates, but the principles can be extended to hydrates as well and to materials that absorb by capillary action rather than by surface reaction. It will be assumed that the moisture is picked up as a surface layer and that this will form a saturated solution. The following notation will be used: The moisture vapor pressure is P_a in the atmosphere, and P' is the vapor pressure of the saturated solution. It follows that P_a must be larger than P' in order that condensation take place (i.e., that moisture is picked up). If $P_a > P'$ then the rate of condensation will be proportional to $P_a - P'$. The larger the surface area, the faster the moisture will condense, so that there is also a proportionality of the rate to the surface area A. This may be expressed

$$\text{Rate} = \gamma = dW/dt = kA(P_a - P') \qquad \text{(III-4-6)}$$

Here k is a proportionality constant and W is the weight of the powder which is adsorbing the moisture. This equation can be integrated to yield

$$W = kA(P_a - P')t + W_0 = \gamma t + W_0 \qquad \text{(III-4-7)}$$

The original weight is here W_0. If a substance is exposed to a relative humidity RH_1, then the rate at first increases linearly with a slope of $kA(P_a - P')$. This is shown in Fig. III-21a. Note that $P' = RH_0(P^*/100)$, where P^* is the vapor pressure of water at the particular temperature and RH_0 is the humidity at one atmosphere of a saturated solution. If such an experiment is performed at several relative humidities, e.g., the set of three denoted RH_1, RH_2, and RH_3 in the figure, it follows from the above

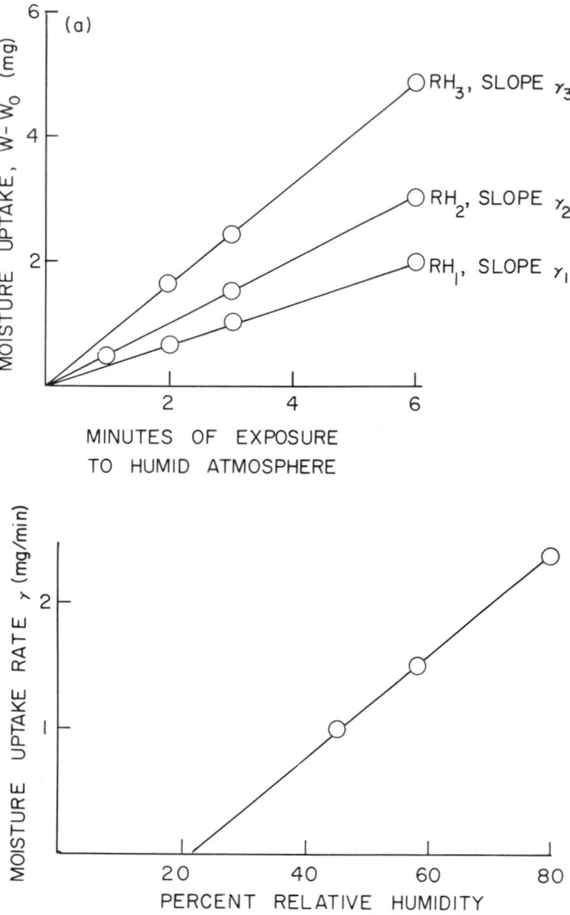

Fig. III-21 (a) Moisture uptake (in milligrams) of a powder, per gram, at three different relative humidities, corresponding to the data in Example III-4-1. (b) Moisture uptake rates of the powder in Fig. III-21a (Example III-4-3).

equation that

$$\gamma = \zeta(\text{RH} - \text{RH}_0) \tag{III-4-8}$$

Note that at an uptake rate of zero Eq. (III-4-8) becomes

$$0 = \zeta(\text{RH} - \text{RH}_0) \tag{III-4-9}$$

In other words, the relative humidity at this point equals that of a saturated solution of the solid in question.

Example III-4-3 The following moisture uptake rates were determined in a separate series of experiments of choline bitartrate:

I (45% RH): (3 mg)/(3 min)
II (58% RH): (4.5 mg)/(3 min)
III (80% RH): (7.2 mg)/(3 min)

Can choline bitartrate be handled (i.e., encapsulated) in an area with a humidity of 35% RH?

Answer It may be seen that the uptake rates are approximately linear functions of the RH (Fig. III-21b):

$$\zeta(\text{I}/\text{II}) = (4.5 - 3)/[(58 - 45)3] = 0.038$$

$$\zeta(\text{II}/\text{III}) = (7.2 - 4.5)/[(80 - 58)3] = 0.041$$

Hence $\gamma - 1 = 0.04(\text{RH} - 45) = 0.04 \times \text{RH} - 1.872$. When $\gamma = 0$, $\text{RH} = 0.872/0.04 = 22\%$. Hence choline bitartrate cannot be encapsulated in an area which has a relative humidity higher than 22%.

REFERENCES

Ashbee, K. H. G. (1968). *In* "Problems in Solid State Physics" (H. J. Goldsmith, ed.), p. 181. Academic Press, New York.
Cahn, D. S., and Fuerstenau, D. W. (1967). *Powder Technol.* **1**, 174.
Cahn, D. S., and Fuerstenau, D. W. (1968). *Powder Technol.* **2**, 215.
Cahn, D. S., Healy, T. W., Fuerstenau, D. W., Hogg, R., and Rose, H. E. (1970). *Nature*, **209**, 499.
Carstensen, J. T., and Anık, S. (1976). *J. Pharm. Sci.* **65**, 158.
Carstensen, J. T., and Patel, M. R. (1977). *Powder Technol.* **17**, 273.
Carstensen, J. T., and Su, K. S. E. (1972). *J. Pharm. Sci.* **61**, 139.
Chien, Y. W., Lambert, H. J., and Grant, D. E. (1974). *J. Pharm. Sci.* **63**, 365.
Chiou, W. L. (1977). *J. Pharm. Sci.* **66**, 969.
Chiou, W. L., and Niazi, S. (1971). *J. Pharm. Sci.* **60**, 1333.
Chiou, W. L., and Riegelman, S. (1971). *J. Pharm. Sci.* **60**, 1281, 1376, 1569.
Cook, P., and Hersey, J. A. (1974). *Powder Technol.* **9**, 257–261.
DeLuca, P., and Lachman, L. (1965). *J. Pharm. Sci.* **54**, 617.
Fessi, H., Marty, J. P., Puisieux, F., and Carstensen, J. T. (1978). *Int. J. Pharmaceutics* **1**, 265.
Garner, W. E. (1955). "The Chemistry of the Solid State," pp. 213–220. Academic Press, New York.
Goldberg, A. H., Gibaldi, M., and Kanig, J. L. (1965). *J. Pharm. Sci.* **54**, 1145.
Guillory, J. K., Huang, S., and Lack, J. (1969). *J. Pharm. Sci.* **58**, 301.
Hersey, J. A. (1975). *Powder Technol.* **11**, 41–44.
Higuchi, T. (1963). *J. Pharm. Sci.* **52**, 1145.
Higuchi, W. I., Mir, N. A., and Desai, S. J. (1965). *J. Pharm. Sci.* **54**, 1405.
Hogg, R., Cahn, D. S., Healy, T. W., and Fuerstenau, D. W. (1966). *Chem. Eng. Sci.* **21**, 1025.

Hogg, R., Mempel, G., and Fuerstenau, D. W. (1968). *Powder Technol.* **2**, 223.
Hollenbeck, R. G., Peck, G. E., and Kildsig, D. O. (1978). *J. Pharm. Sci.* **67**, 1599.
Johnson, M. C. R. (1972). *Pharm. Acta Helv.* **47**, 546.
Kristensen, H. G. (1973). *Powder Technol.* **7**, 249.
Lacey, P. M. C. (1943). *Trans. Inst. Chem. Eng.* **21**, 53.
Lacey, P. M. C. (1954). *J. Appl. Chem.* **4**, 257.
Marshall, K., and Sixsmith, D. (1974/1975). *Drug Dev. and Ind. Pharm.* **1**, 51.
Mering, J. (1948). *Colloq. Int. C.N.R.S.* **10**, 189.
Nakai, Y., Fukuoka, E., Nakajima, S., and Yamamoto, K. (1977). *Chem. Pharm. Bull.* **25**, 3340.
Nakai, Y., Fukuoka, E., Nakajima, S. I., and Iida, Y. (1978a). *Chem. Pharm. Bull.* **26**, 2983.
Nakai, Y., Nakajima, S. I., Yamamoto, K., Terada, K., and Konno, T. (1978b). *Chem. Pharm. Bull.* **26**, 3419.
Nicholson, W. J., and Smith, J. C. (1966). *Chem. Eng. Prog.* **62**, 83.
Orr, N. A., and Shotton, E. (1973). *The Chemical Engineer* **269**, 12.
Poole, K. R., Taylor, R. F., and Wall, G. P. (1964). *Trans. Inst. Chem. Eng.* **42**, T305.
Ridgway, K., and Segovia, E. (1968). *J. Pharm. Pharmacol. Suppl.* **20**, 194S–203S.
Rippie, E. G., Olsen, J. L., and Faiman, M. (1964). *J. Pharm. Sci.* **53**, 1360.
Rippie, E. G., Faiman, M., and Pramode, M. K. (1967). *J. Pharm. Sci.* **56**, 1523.
Roseman, T. J. (1975). *J. Pharm. Sci.* **64**, 1721.
Roseman, T. J., and Higuchi, W. I. (1970). *J. Pharm. Sci.* **59**, 353.
Sekiguchi, K., and Obi, N. (1961). *Chem. Pharm. Bull.* **9**, 866.
Sekiguchi, K., Ueda, Y., and Nakamori, Y. (1963). *Chem. Pharm. Bull.* **11**, 1108, 1123.
Sekiguchi, K., Obi, N., and Ueda, Y. (1964). *Chem. Pharm. Bull.* **12**, 134.
Shah, S. A., and Parrott, E. L. (1976). *J. Pharm. Sci.* **65**, 1783.
Shefter, E., and Kmack, G. (1967). *J. Pharm. Sci.* **56**, 1028.
Stange, K. (1954). *Chem. Ing. Tech.* **26**, 331.
Stange, K. (1963). *Chem. Eng. Tech.* **35**, 580.
Travers, D. N. (1975). *Powder Technol.* **12**, 189–190.
Zhdanov, G. S. (1965). "Crystal Physics." Academic Press, New York.

CHAPTER

IV

Precompression Operations

In order to make a tablet out of a powder, one must impart the following qualities to it:

(a) The powder must be compressible, i.e., upon application of a force the individual particles must bond together.

(b) The particles must flow well. If they do not, then tablet uniformity will not be assured.

(c) The surfaces (particularly of the drug substance) must wet well. If not, then the rapid dissolution of the active ingredient is not insured.

If a compound has all these properties, then it may simply be blended with excipients and compressed. This is called direct compression and will be dealt with in one of the subsequent chapters. If it contains all the above properties except (b), i.e., it does not flow well, then the material can be precompacted, either by slugging or chilsonating. If the powder neither flows well nor is self-compressible, it is usually wet granulated, and this process will be dealt with first.

IV-1 WET GRANULATION

Wet granulation is a procedure by which (a) compressibility is improved by the addition of a binder, (b) flowability is attained at the same time due to a particle size increase (refer to Section II-9), and (c) the

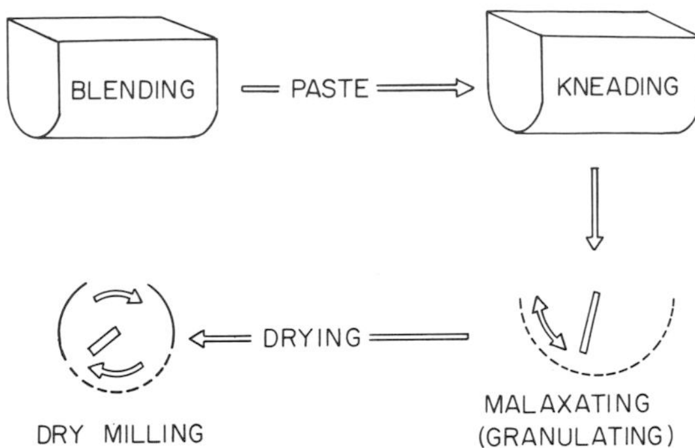

Fig. IV-1 Flow chart of wet granulation.

hydrophobic surfaces are made hydrophilic by use of a hydrophilic binder. Prior to discussing the individual ingredients it is worthwhile looking at the process of wet granulation. The steps in granulation are as shown in Fig. IV-1.

(1) blending (the aspects of blending have already been discussed in Section III-3),
(2) addition of a solution of the binder (paste),
(3) kneading,
(4) malaxation (granulation),
(5) drying, and
(6) milling.

These will be discussed one by one.

IV-2 PASTE AND BINDERS

The quality of a good binder is that it (as the name implies) causes particles to bind together. It is desirable that the cohesion C imparted by the binder (paste) be large, and that the viscosity η be as small as possible. Second, it should be inert. Third, it should blend easily. It should be as soluble as possible (in the cold), so that it can be made with a minimum of solvent (water). It should be as nonhygroscopic as possible.

In wet granulation it is used directly as a paste or it is blended in with the powders and a solvent (water) is added. Alternatively, a concentrated paste can be used with additional addition of solvent (water) to finish the granulation. The different types of binders used in present day pharmaceutical manufacture are

(1) the starches,

(a) cornstarch and wheat starch in the United States and
(b) potato starch in Europe,

(2) gelatin,
(3) povidone,
(4) sugars,
(5) gums, and
(6) cellulose derivatives.

The use of a binder is exemplified in the manner in which a starch paste is made. Starch is usually used in a concentration of 5–10% in water, and it is made by taking a mass x of starch and suspending this in a mass x of cold water. For a 1:10 starch paste one adds this suspension to a volume $9x$ of boiling water, or one adds $9x$ of boiling water to the suspension. The two modes of addition cause gels of different consistency, as shown in Fig. IV-2. The starch particles in the former case experience higher temperatures than in the latter case, and therefore hydrolize more. The mechanisms of bonding are (Rumpf, 1958)

(1) solid bridges,
(2) forces that are of an interfacial or capillary nature,
(3) adhesion and cohesion,
(4) attractive forces, and
(5) mechanical forces.

According to Newitt and Conway (1958) there are four particular states of granular formations: the pendular state, the funicular state, the capillary

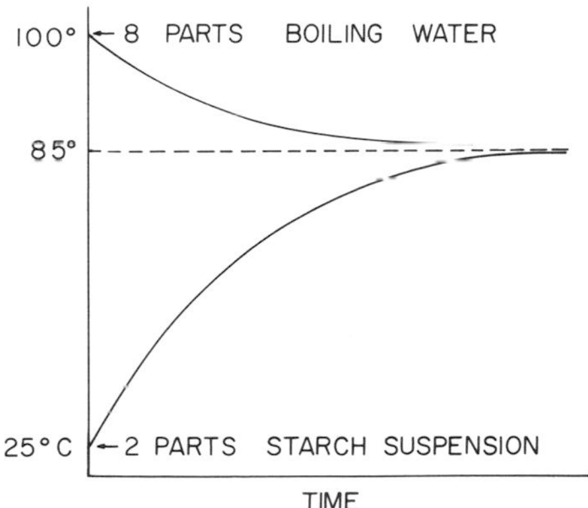

Fig. IV-2 Temperature experienced by the starch granule according to whether the suspension is added to boiling water or boiling water is added to the starch suspension.

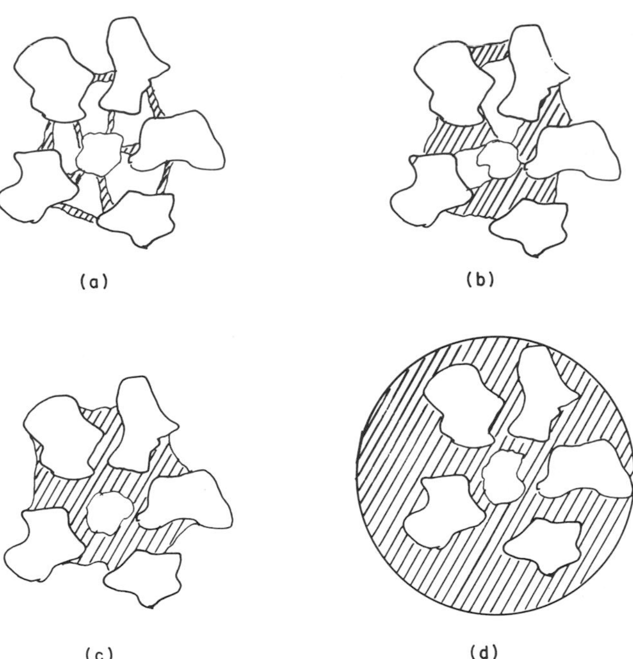

Fig. IV-3 Types of wet granules: (a) pendular, (b) funicular, (c) capillary, and (d) droplet. [After Newitt and Conway (1958).]

state, and the droplet state (Fig. IV-3). Aside from the three mentioned properties (size, flow, and hydrophilization), the granulation process, therefore, also adds the parameter of porosity to the ensuing particle (granule).

The principal parameters of the granulation (Cruaud, 1979) are the concentration of the binder in the paste, the volume and temperature of the solution, the time of kneading (Zoglio *et al.*, 1976; Carstensen *et al.*, 1976), the pressure in malaxation, and the milling step. In general the properties of a binder which are studied are the following. One studies a *solution* as far as (a) rheological behavior (i.e., whether it is pseudoplastic, thixotropic, or rigid), (b) the surface tension and the contact angle, and (c) the concentration and the molecular weight (when a series of binders is studied). The second point of study is the *sieve analysis*: (a) the type of distribution curve one obtains (normal, log-normal, Weibull, or bimodal) (Steiner *et al.*, 1974) and (b) the percentage of fines ($< 200\ \mu m$). The third point one studies is the cohesion of the granules (Veillard, 1976): (a) first the friability (Shafer *et al.*, 1956), then (b) the hardness of the granule (Harwood and Pilpel, 1968). In both of these measures there is a correlation between the hardness and the percent of granules which remain at the

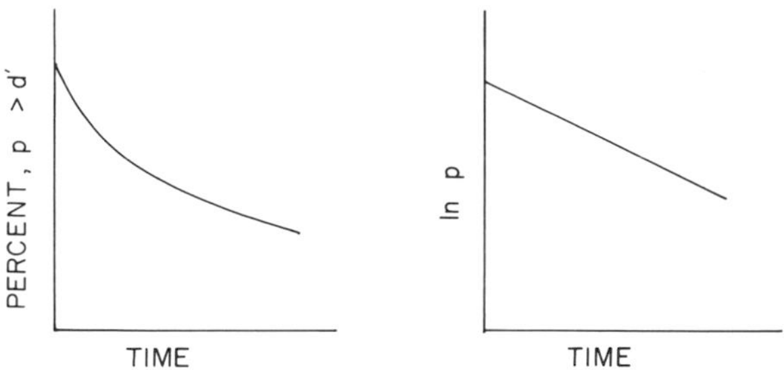

Fig. IV-4 Amount of material of sieve size d', remaining on a sieve of size d' as a function of time of milling or friabilization. A fine powder of diameter d'' is formed during the milling, causing the amount of particles of the original diameter to diminish.

original diameter (Carstensen *et al.*, 1978; Mehta *et al.*, 1977). Figure IV-4 shows a typical curve of percent larger than or equal to original diameter versus milling time and shows how this can be used to calculate the binding force, since this latter is a function of the logarithmic relation between percentage above original diameter and time.

Another parameter which is useful in the study of granulation is the apparent volume of the final granulation and the *porosity* (Duchene, 1976). Binders and pastes can also be studied by means of films cast from the binder material (Healey *et al.*, 1974). Table IV-1 shows the breaking strength ($J\,cm^{-2}$) of some common binders. Note that this breaking strength is a function of the moisture content of the material, and this is one of the reasons that drying, as will be discussed a little later, is very important in granulation technology.

A further point of study is the distribution of the active component. Obviously granulation should impart uniformity to the solid mixture, and in general wet granulated materials show a smaller content standard

TABLE IV-1

Breaking Strength of Binders[a]

Binder	Moisture content of film (%)	Breaking strength ($J\,cm^{-2}$)
Gum arabic	9.8	1.4
Gelatin	10.8	12
	13.5	7.2
Methylhydroxyethylcellulose	3.1	34
Povidone	10.4	1.0
Starch	8.1	18

[a] Healey *et al.*, 1974.

deviation than simply blended dry powders. The next two points of study will be dealt with somewhat later, but obviously the drying rate of the granulations and their compressibility are of importance in the selection of a binder and a method of manufacture.

IV-3 KNEADING

The length of kneading time in the kneading step is a very important factor in granulation. In essence the steps that occur in kneading (Zoglio *et al.*, 1976; Carstensen *et al.*, 1976) are

(a) an excessive wetting, which is localized (Fig. IV-5a),
(b) distribution of the paste (Fig. IV-5b),
(c) formation of an equilibrium granule (usually at an optimum kneading time of 5–10 min) (Fig. IV-5c), and
(d) in the case of a soluble binder, excipient, or active ingredient, excessive dissolution of material into the liquid, densification (decrease in porosity), and an increase in particle size (Fig. IV-5d), caused by further kneading.

It has been shown that step (d) will cause the following to happen: Aside from increase in particle size, it also causes a higher flow rate initially (because the particles become smoother). This will later decrease again because of twinning of the larger particles. It causes a slowing of the drying of the granules (because of the decreased porosity), and it causes a decrease in dissolution time.

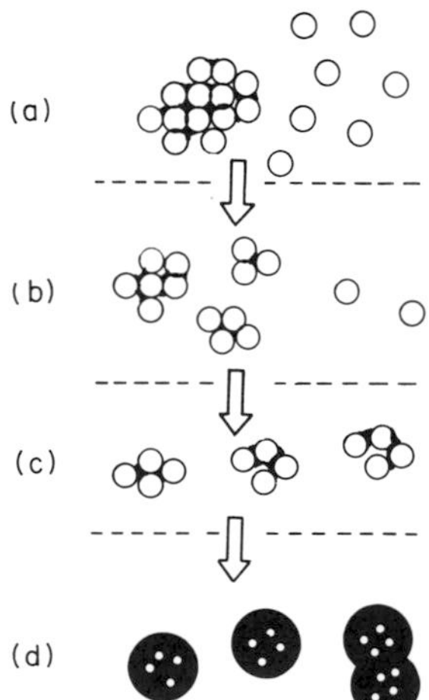

Fig. IV-5 Stages during the wetting of a powder with a paste: (a) excessive, localized wetting plus nonwetted powder, (b) solvent or paste distribution, (c) equilibrium granules, and (d) overkneaded granule.

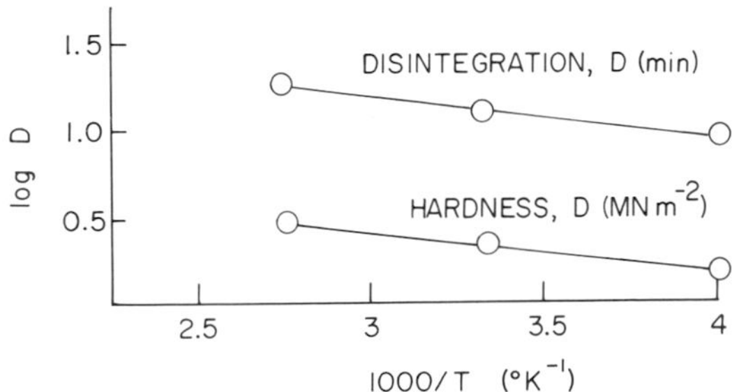

Fig. IV-6 Effect of the temperature of the starch paste at the time of granule formation on the hardness and disintegration of the tablets finally produced. [After Pilpel and Esezobo (1977).]

The temperature of kneading is also important [see Pilpel and Esezobo (1977), and Fig. IV-6]. As seen, both disintegration and hardness of tablets that are produced from granules increase with increasing temperature and follow a type of Arrhenius relationship.

The distribution of the binder paste has been investigated by Dingwall and Ismail (1977). They granulated glass beads and found that in general less binder (paste) was distributed to the larger bead. However, only fairly coarse beads were tested and the study is not conclusive, although very instructive. It was found that the distribution was fairly regular with povidone and gelatin granulations but was rather irregular with methocel and starch pastes.

The binder can also migrate during drying, as pointed out by Rubinstein and Ridgway (1974). Figure IV-7 shows the amount of povidone on the surface of the granule and in the core as a function of the final moisture content of the granule.

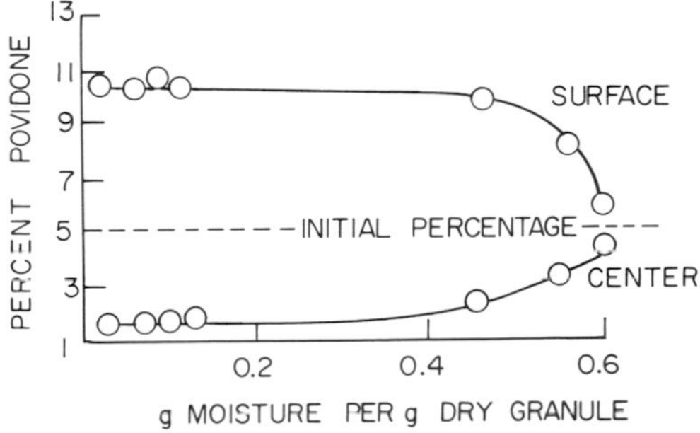

Fig. IV-7 Migration of binder from core to surface of the granules as a function of moisture content of the granules. [After Rubinstein and Ridgway (1974).]

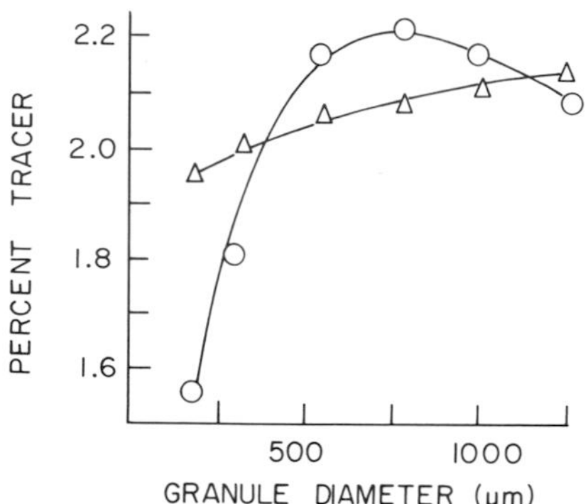

Fig. IV-8 Distribution of the active ingredient (tracer) as a function of the amount of granulation fluid (volume per volume of granulation). △, 14%; ○, 16%. [After Selkirk (1976).]

The amount of liquid that is used for the granulation is also important in respect to the distribution of paste as a function of particle size of the granule. Selkirk (1976) has shown this and the results are shown in Fig. IV-8. There is, of course, an optimum amount of binder. Carstensen *et al.* (1978) have shown that in attrition of granules, the diameter of the fine fraction (d'') is a measure of the quality of the granulation; i.e., the larger d'' the better the granulation. By plotting d'' as a function of percentage of binder (povidone), they could arrive at the optimum amount of povidone in a granulation (Fig. IV-9).

Note that the choice of binder can be important so far as dissolution of

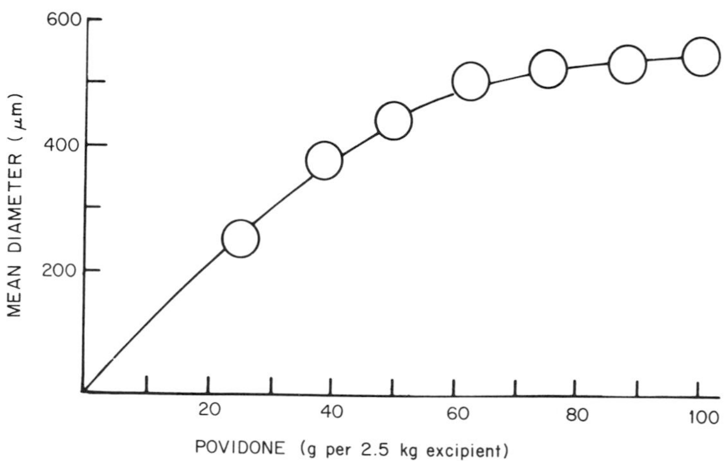

Fig. IV-9 Diameter of the fine fraction (ordinate) produced during milling of a povidone granulation as a function of the percent of povidone used. [After Carstensen *et al.* (1978).]

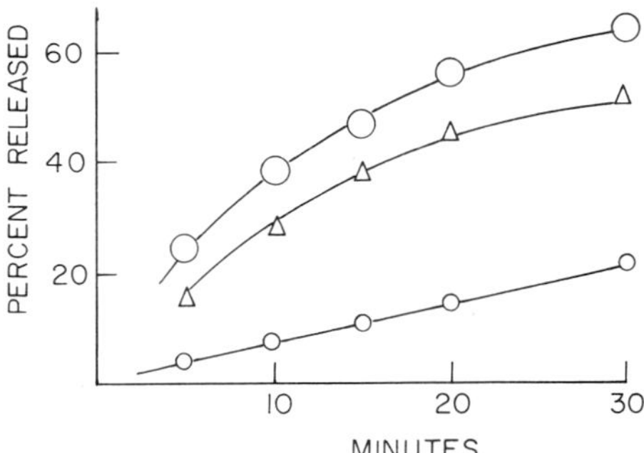

Fig. IV-10 Effect of various binders on dissolution rate. Top curve (large circles), $Pharmagel_B$ (wet granulated); middle curve (triangles), microcrystalline cellulose (direct compression); lower curve (small circles); Kollidon (wet granulated). [After Jaminet *et al.* (1969).]

the active ingredient is concerned. Figure IV-10 shows the influence of various binders on dissolution (Jaminet *et al.*, 1969).

Finally, it should be mentioned that it is *never* possible to scale up the amount of granulation liquid used. This is due to several factors; e.g., the kneading action is usually stronger in a large kneader, and hence the granules are more compact, the "drying out" during the kneading is different (different surface geometries), the amount "lost" to the wall (in percent) is less in a large kneader, and the temperature of the binder gel is obviously different. In any event, the amount of water needed in large batches is usually *less* than in a small blender on a percentage basis. For the first large scale batch produced of a product it is therefore customary to make a paste 10% more concentrated than for the small batch. After addition of all the paste, sufficient water is added to complete the fluid addition step. (This can be gauged, for instance, by eye or by wattage in the kneader.) This added water is less than 10% of the deficiency of water in the paste and allows recalculation of the paste composition for the next (and subsequent) batches.

IV-4 DRYING OF GRANULATIONS

In the fine-chemicals industries drying is carried out to remove solvent and to arrive at a pure substance, but in pharmaceutical operations, such as the ones we are dealing with, drying is primarily confined to granulations. In drying wet granules, there will be a mass x of water which has been added to 1 kg of dry weight of the granulation. The drying curve (as defined here) will be a plot of x versus the time of exposure of the granulation to the drying air stream.

IV-4-1 The Psychrometric Chart

Very rarely is an air sample completely anhydrous, so to have a basis for comparison it is conventional to express moisture contents in air as *absolute humidity* Y. This denotes the number of kilograms of water per kilogram of dry air. In other words the lower the absolute humidity of an air sample (at a given temperature), the faster it will dry a given substance (granulation).

Humidity is also measured as *relative humidity* (RH). This is 100 times the ratio between partial water vapor pressure P and saturation water vapor pressure P' of the air sample at the particular temperature in question. For a given air sample the partial vapor pressure of the water and of the air will add up to the total (measured) pressure (which in most cases is 1 atm = 101 kPa†), i.e.,

$$P_{\text{water}} + P_{\text{air}} = P_{\text{total}}. \qquad \text{(IV-4-1)}$$

The number of moles of water and air can be calculated from the ideal gas law

$$n_{\text{water}} = P_{\text{water}}V/RT \qquad \text{(IV-4-2)}$$

where n is the number of moles, V is the volume in cubic meters, R is the gas constant (8.3 J/mole °K) and T is the absolute temperature. If the total pressure is 101 kPa, then $P_{\text{air}} = 101 - P_{\text{water}}$, so that

$$n_{\text{air}} = (101 - P_{\text{water}})V/RT \qquad \text{(IV-4-3)}$$

It is now possible to obtain the amount of the constituents present in 1.0 m^3 of moist air by simply knowing what the water vapor pressure is.

Example IV-4 What is the mass (M_{air} and M_{water}) of water and air per cubic meter of air sample in saturated air at 50°C (323.15°K)?

Answer The water vapor pressure at 50°C is 12.3 kPa, so $P_{\text{air}} = 101 - 12.3 = 88.7$ kPa. The molecular weights are 0.018 kg/mole for water 0.029 kg/mole for air. Hence

$$M_{\text{water}} = 12.3 \times 10^3 \times 0.018/(8.3 \times 323.15) = 0.0817$$

$$(\text{kg water/m}^3 \text{ air sample}) \qquad \text{(IV-4-4)}$$

$$M_{\text{air}} = 88.7 \times 10^3 \times 0.029/(8.3 \times 323.15) = 0.9496$$

$$(\text{kg air/m}^3 \text{ air sample}) \qquad \text{(IV-4-5)}$$

† For unit conversions see Appendix.

The absolute humidity in the example is 0.0817/0.9496 = 0.086 kg of water per kilogram of dry air. The value of $R = 8.3$ J mole^{-1} deg^{-1} has been used in this calculation. One might assume that for 50% relative humidity the absolute humidity would be about one half of this figure. This is not quite true as shown in the calculation below:

$$M_{\text{water}} = 0.0408 \quad \text{(kg water)/(m}^3 \text{ air sample)}$$

$$M_{\text{air}} = 1.015 \quad \text{(kg water)/(m}^3 \text{ air sample),}$$

i.e.,

$$Y = 0.0408/1.015 = 0.0402 \quad \text{(kg water)/(kg dry air)}$$

Figures like this can be found directly using a psychrometric chart (Fig. IV-11). In fact, the curves that are shown in the psychrometric chart are calculated in the fashion shown above. Because most charts are presently available in old units, Fig. IV-11 is also shown in old units, but the conversion factors are shown in the legend below the figure. To use the chart for the situation above, the temperature (50°C, 120°F) is located on the horizontal axis. The perpendicular is drawn, the intersection with the curve labeled 50% is located, and the ordinate value (corresponding to the Y axis on the right) is determined (0.04 kg of water per kilogram of dry air).

Densities and relative humidities of moist air at particular temperatures can be important as well. In this case, the line labeled "SP. vol. dry air" (with an ordinate on the humid volume scale) is used. Next, the value indicated by the curve "Sat'd humid volume" is found, and the prorated average between the two (in the case of 50%, simply the average) is determined. The values are (14.5 ft^3/lb =) 0.086 m^3/kg and (16.5 ft^3/lb =) 0.98 m^3/kg, so the average value is 0.92 m^3/kg. In the example $M_{\text{total}} = 0.92 + 0.08 = 1.0$ kg/m^3, i.e., 1.0 m^3/kg.

Drying is evaporation of water (solvent), and this is accomplished by adding heat Q (in joules) to the powder or granulation. Denoting by L^* (J/kg), the latent heat of vaporization of water, the mass (kg) which can be evaporated would be L^*/Q. It is, therefore, important to have knowledge of the heat content H (in joules per kilogram of dry air) or enthalpy of the air. One can, in this manner, calculate the heat Q given off to the granulation as the difference between the total enthalpy value of the incoming (H_0) air and that of the outgoing (H_1) air. Heat contents of in- and outgoing air samples can be obtained from the psychrometric chart in the same way as the humid volumes: At, for instance, 50% RH and 50°C (122°F), a vertical line is drawn at 122°F, and the intersection with the line denoted "H of dry air" (20 BTU/lb dry air at the left scale) is noted. Next the intersection with the line "H sat'd air" is noted (110 BTU) so that the

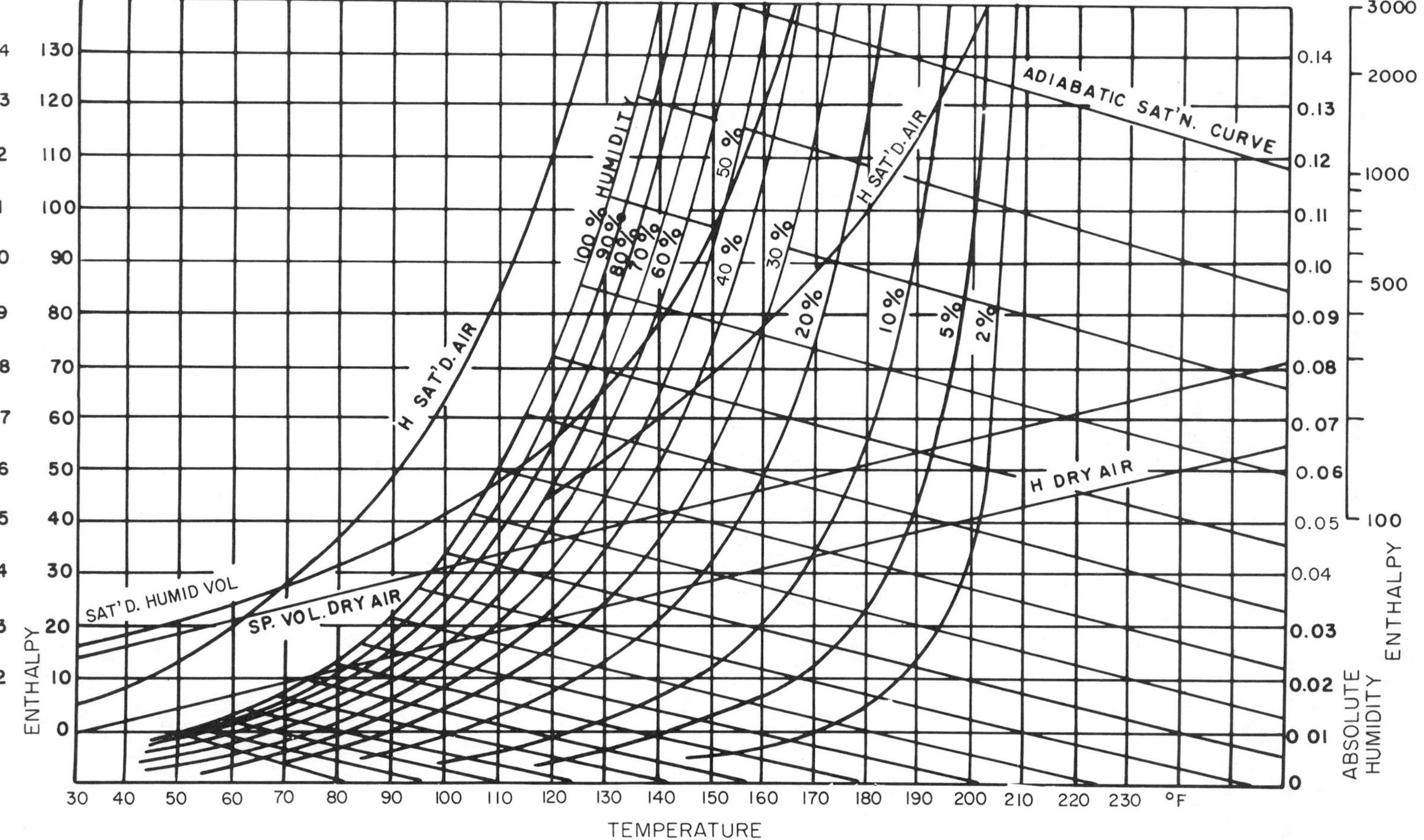

Fig. IV-11 Psychrometric chart. The units given are BTU, °F, cubic feet and pounds. These can be converted to new international units (Appendix II) as follows: To convert from °F to °C subtract 32 from °F and divide by 1.8; to convert from ft^3 to m^3 multiply ft^3 by 0.027; to convert from BTU to J multiply BTU by 1055; to convert from lb to kg, divide lb by 2.2.

enthalpy of a 50% sample is the average of these two figures:

$$(110 + 20)/2 = 65 \text{ BTU/lb} = 65 \times 1055 \times 2.2 = 151{,}000 \text{ J/kg}$$
$$= 151 \text{ kJ/(kg dry air)} \qquad \text{(IV-4-6)}$$

The psychrometric chart, of course, can be used to determine RH values once wet and dry bulb temperatures are known. These are usually obtained by means of a sling psychrometer. If, for instance they were 40°C (104°F) and 50°C (120°F), respectively, then a vertical line is drawn at 104°F to the 100% humidity curve. The downward slanted line (adiabatic saturation curve) is then followed down until it crosses a vertical line through 120°F. This is on the curve marked 50% RH.

In a drying operation the air will enter the dryer at a certain rate W (kg dry air/sec), and the same amount of *dry* air leaves the dryer. A larger amount of air on a moist basis leaves than enters the dryer because the exit air contains whatever water was evaporated during the time the air spent in the dryer. By using a psychrometric chart and calculating as above, one can, by knowledge of flow rate of air and drying time t (sec), arrive at the amount M_1 (kg) of water which has evaporated. This, then, should be equal to the amount of water lost by the granulation and determined by an appropriate moisture assay before and after (M_2 kilograms of water lost). In reality M_1 never equals M_2, and the ratio between the two is a measure of how efficient the dryer is.

Aside from the mass and heat balances which have just been discussed, the kinetics of drying, i.e., the rate, is exceedingly important and will be the subject of the following discussion.

IV-4-2 Kinetics of Drying

The efficiency of a dryer is a thermodynamic efficiency, but how rapidly a dryer will dry a granulation is, of course, also of importance. The rate of drying is usually a function of the surface A (m^2) and the load L (kg dry weight). The proportionality constant N is a type of heat transfer coefficient. It is necessary to distinguish between moisture that is bound chemically (such as water of hydration of a salt hydrate) and the water which is not chemically bound.

In this chapter the nomenclature used will be as follows. A *dry solid* includes the *bound* moisture. The symbol L is used for dry weight (kg), the anhydrous weight (denoted L') plus the weight of bound water. Since X is the symbol used for kilograms of water per kilogram of dry solid, then the mass m of unbound water (kg) is equal to LX.

There are three processes involved in the drying of granules (Cartensen, 1973). The water is present either on the surface of the granule, or in the (porous) void space of the interior of the granule, or it is chemically bound. Obviously the surface moisture dries first, and this occurs at a constant rate and (Shepherd *et al.*, 1938; Lewis, 1921) during this so-called *constant rate period*,

$$dm/dt = L(dX/dt) = -NA \quad \text{(IV-4-7)}$$

$$m = m_0 - NAt \quad \text{(IV-4-8)}$$

Here N is the apparent heat transmission coefficient and m_0 is the moisture initially present. The drying curve is shown in Fig. IV-12, and the part AB is the constant rate period.

Once all the moisture on the surface of the granules is removed, the second phase, i.e., removal of the internal moisture, starts. This is a diffusional process (Sherwood, 1931; Newman, 1931, pp. 203, 310; Sherwood and Comings, 1933). It is a *falling rate* period where the rate is a linear function of X:

$$dm/dt = L\,dX/dt = -AN(bX + a) \qquad (t > t_c, X > X_c) \quad \text{(IV-4-9)}$$

as shown in Fig. IV-12, and X_c and t_c are called critical moisture content and critical time. The equation can be integrated to read

$$\ln[(X + a/b)/(X_c + a/b)] = -(NbA/L)(t - t_c) \quad \text{(IV-4-10)}$$

Since X is the moisture which is not bound, the curve will be upwards concave, as shown in Fig. IV-12 (segment BC). Frequently a is of small magnitude, so that if one denotes by X^* the total moisture content (i.e., that obtained by moisture assay), then Eq. (IV-4-10) takes the form

$$\ln[(X^* - X_b)/(X_c - X_b)] = -(NbA/L)(t - t_c) \quad \text{(IV-4-11)}$$

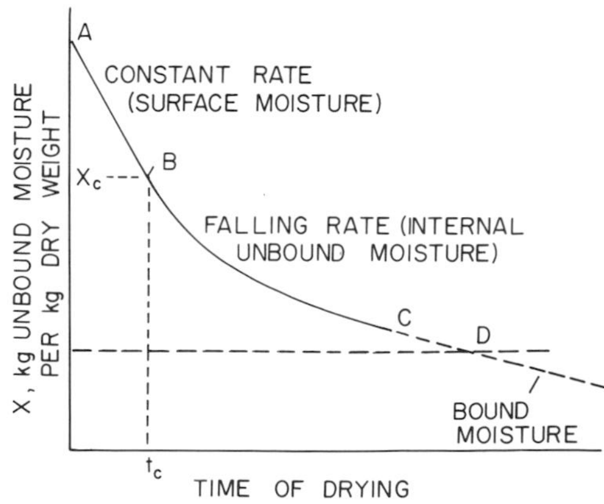

Fig. IV-12 Drying curve showing kilograms of water per kilogram of dry solid (X) as a function of drying time.

The bound moisture X_b can be calculated as the asymptotic value towards which the quantity X^* tends during the falling rate (Fig. IV-12). If drying is carried out to a lower moisture content, then the structure of the granule is frequently destroyed, and this affects the machinability of the granule. The practice of overdrying followed by addition of a certain amount of moisture to get a granulation that meets a certain set of specifications is, therefore, not a good procedure.

Internal porosity frequently affects the falling rate. But there are other factors. All have one thing in common, that diffusion is the rate-determining step (Oliver and Newitt, 1949; Pearse *et al.*, 1949; Nissan *et al.*, 1960; Bell and Nissan, 1959; Adams, 1962).

IV-4-3 Pharmaceutical Methods of Drying

The most common methods of drying in pharmaceutical practice are tray, vacuum, and fluid bed drying. Less common but also used are rotary countercurrent, truck, and tunnel drying. There are also two methods that are often used in the chemical industry (drum and spray drying) and are practically never used in the pharmaceutical industry.

The various operations germane to pharmaceutical operation (in ascending order of importance) will be described briefly in the next section.

IV-4-4 Truck Dryers

In truck drying, the material to be dried is placed on trays. The trays are placed on trucks, which in turn are placed in a room. Gentle movement of drying air across the surface of the tray causes the drying of the substance in the tray. Truck drying is used for drying of soft shell capsules. In this case the drying air has a temperature of 37°C and a relative humidity of less than 10%. The used air is redried by passing it over silica or through saturated LiCl solutions. The drying of the soft shell capsule is a diffusion process which is modelistically akin to diffusion from a cylinder. This case is described by Jost (1960). The equation is

$$\ln(c - c_\infty) = -t/\Gamma + \ln(c_0 - c_\infty) \qquad \text{(IV-4-12)}$$

$$\Gamma = h^2/5.8d \qquad \text{(IV-4-13)}$$

Here c is the amount of moisture at a time t, and c_0 and c_∞ imply initial and equilibrium moisture content of the capsule. The thickness of the gelatin film is denoted h, and the gelatin film has a diffusion coefficient of D, which is presumed independent of moisture content. Equation (IV-4-13) defines Γ. Actual experience with drying of soft shell capsules adheres well to Eq. (IV-4-12). The drying end interval is very critical. Overdrying causes brittleness of the capsule, and insufficient drying gives an excessively soft and sticky capsule which will deform when stored.

IV-4-5 Tunnel Drying

Tunnel drying is not a common method in the pharmaceutical industry in terms of the number of installations. It should be pointed out, however, that the few times when it is found the operations are fairly large scale. A couple of examples are hard shells for hard shell capsules, hard candy products (pharyngets), and carrier-type diagnostic aids (filter paper). In these drying operations the wet material enters the tunnel where it moves slowly. Drying takes place gradually in such a fashion that the unit upon exit has the desired moisture content. Heat is derived from either hot air or infrared light. The mechanism of drying and the drying equation that results from it are of the same nature as those which will be derived in the next section (countercurrent drying).

IV-4-6 Countercurrent Drying

Rotary dryers are based on countercurrent drying. They are long cylinders with baffle arrangements internally (sometimes in a helical fashion). This directs the product in a direction which is counter to the flow of air. The dryer rotates so that the granules will cascade through the air all the time, and at the point where the air is dry the product is also fairly dry so that the nature of the flow is indeed countercurrent. This type of drying is usually only used for large volume products and usually entails a semiautomated process (Fig. IV-13). Pitkin and Carstensen (1973) have demonstrated that the rate-limiting step in this type of operation is the movement of moisture inside the granule. They showed that for such a case

$$\frac{c - c_\infty}{c_0 - c_\infty} = \frac{6}{\pi^2} \sum_{j=1}^{\infty} \frac{1}{j^2} \exp\left(-\frac{jt^2}{K}\right) \tag{IV-4-14}$$

$$K = a/4\pi^2 D \tag{IV-4-15}$$

where j is a running index, a is the granule diameter, and D is the water diffusion coefficient inside the granule. This latter is temperature dependent:

$$D = D_0 \exp(-E'/RT) \tag{IV-4-16}$$

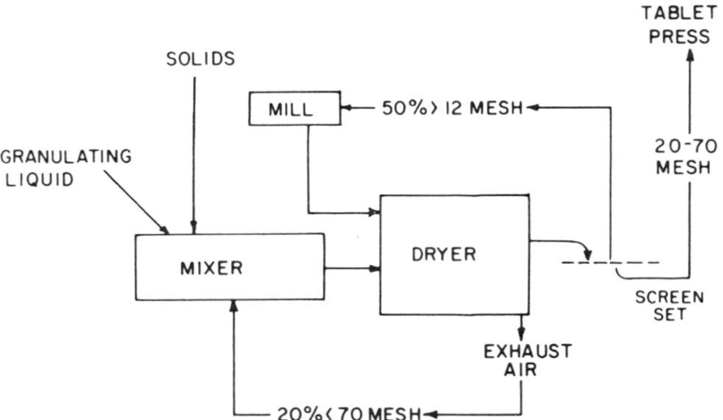

Fig. IV-13 Counter drying set-up in a semiautomated granulation sequence. [Courtesy of Lewis-Howe Co., St. Louis, Missouri.]

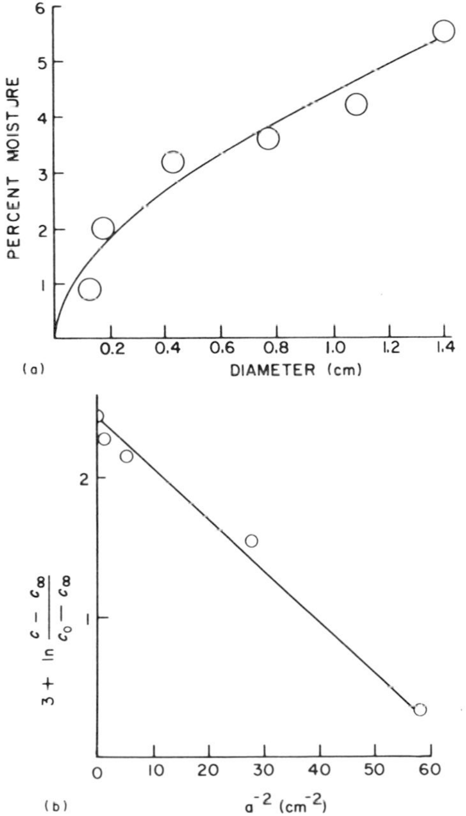

Fig. IV-14 (a) Moisture content as a function of particle size and (b) according to Eq. (IV-4-17). [After Pitkin and Carstensen (1973). Reproduced with permission of the copyright owner, the *Journal of Pharmaceutical Sciences*.]

Here E' is an activation energy. Since (Fig. IV-13) a range of particle sizes exit from the dryer, Eq. (IV-4-14) will predict that the moisture content is a function of the particle diameter, because the drying time t' is the same for all of the exiting particles. In this case (realistic values of t') terms with j larger than unity disappear from Eq. (IV-4-14), which then becomes

$$\ln[(c - c_\infty)/(c_0 - c_\infty)] = -t'4\pi^2 D/a^2 \qquad \text{(IV-4-17)}$$

That means that the logarithm of $(c - c_\infty)/(c_0 - c_\infty)$ is a linear function of a^{-2}, and Fig. IV-14 shows the validity of this. The diffusion coefficient can be obtained from the slope and is 3×10^{-11} m^2/sec, i.e., of the correct order of magnitude. The intercept is, however, not zero, presumably because of entry effects (surface moisture).

IV-4-7 Tray Drying

In tray drying the granulation is placed on a tray, which in turn is placed in an oven. The tray can also be placed on trucks, which in turn are placed in *individual* ovens (as opposed to truck drying, in which many trucks are placed in one room). The air comes in from one wall of the oven, then passes over the surface of the trays and exits at the opposite wall. Tray drying is not in a pioneering area of technology (since it is both slow and inefficient), but it is widely used and because of capital investments it is not apt to be replaced entirely.

There have been several reports in literature on the subject of drying from trays (Ridgway and Callow, 1967; Luikov, 1963; Morgan and Yerazunis, 1967; Bhutani and Bhatia, 1975; Opankunle *et al.*, 1975; Gilliland, 1938). If the tray is filled to a depth (in meters) of a with wet granulation and if the rate limiting step is moisture transfer from bed to air, then (Jost, 1960) the loss rate will follow the equation

$$-D\,\partial C/\partial x = \zeta(C_0 - C_s) \qquad \text{(IV-4-18)}$$

where D is the diffusion coefficient (m^2/sec) of water in vapor form, $\partial C/\partial x$ (kg/m^4) is the gradient of moisture (in the vapor phase) over the bed-to-air interface, subscripts s and 0 represent air and bed, respectively, C is the concentration (kg/m^3) of water in vapor form in the void space of the bed, and ζ is a proportionality constant.

The assumption is made that the air is dry, and therefore $C_s = 0$. Therefore $C_0 - C_s$ will be denoted C in the following. When, initially, there is surface moisture present, the vapor in the void space has a vapor pressure equal the saturation pressure, P_{sat} ($N\,m^{-2}$), i.e.,

$$C = P_{sat}0.018/RT \quad kg/m^3 \qquad \text{(IV-4-19)}$$

Therefore C is constant, and applying Fick's law to this period gives the drying rate as dm/dt (kg/sec):

$$-(1/\epsilon A)(dm/dt) = D(\partial C/\partial x) = \zeta C \qquad \text{(IV-4-20)}$$

where A is the tray surface and the porosity of the bed is ϵ. Therefore $A\epsilon$ is the cross section for diffusion. By introducing Eqs. (IV-4-18) and (IV-4-19) into Eq. (IV-4-20), the zero-order rate of evaporation is obtained:

$$-(dm/dt) = (A\epsilon\zeta)0.018P_{\text{sat}}/RT \qquad \text{(IV-4-21)}$$

It is therefore possible to calculate the constant ζ from the negative slope of this initial drying curve, provided the granules contain surface moisture. It should be pointed out that for properly processed granules this is often *not* the case.

At a particular point in time the surface moisture will have been removed, and from this point on the water vapor pressure in the interparticulate space will be less than P_{sat}. If now both the internal diffusion of water in the granule and the evaporation from granule surface to gas phase are rapid, then the transfer of moisture over the interface between bed and stream will be rate controlling. The diffusion equation can now be solved (Crank, 1970, p. 56). The parameter H [in Eq. (IV-4-22)] is introduced (a is the thickness of the bed and q serves to make H dimensionless):

$$H = aq/D \qquad \text{(IV-4-22)}$$

(i.e., the dimensions of q are centimeters per second). Introducing this into the approximation solution to Eq. (IV-4-18) gives

$$\begin{aligned} \ln(1 - c/c_\infty) &= -(\beta^2 D/a^2)t + \ln\left[2H^2/\beta^2(\beta^2 + H^2 + H)\right] \\ &= -Gt + \ln K \end{aligned} \qquad \text{(IV-4-23)}$$

where $-G$ and $\ln K$ are the respective slopes and intercepts of the log-linear equation, and where β is the smallest positive root obtained from the equation

$$H = \beta \tan \beta \qquad \text{(IV-4-24)}$$

Here c is the mass of water (in kilograms) present in the granules at time t and c_∞ is the equilibrium amount of moisture in the granulation, i.e., that obtained by proper drying of the product, i.e., after removal of unbound moisture.

Equation (IV-4-23) implies that the amount of moisture beyond the equilibrium moisture is log-linear in time, and that the quantity β^2 can be obtained from the value of the slope (i.e., $-G$):

$$\beta^2 = Ga^2/D \qquad \text{(IV-4-25)}$$

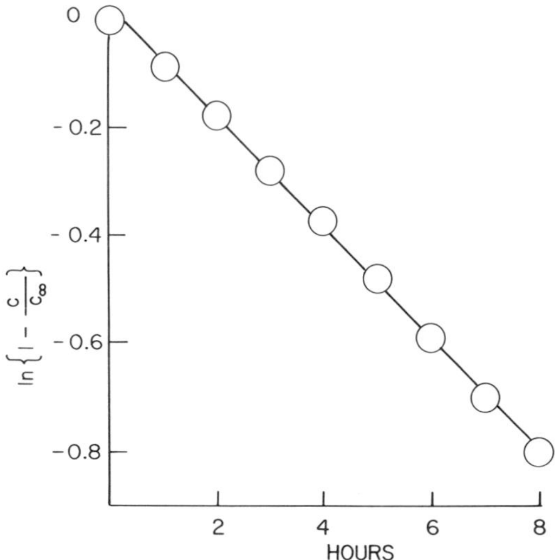

Fig. IV-15 Drying curves for a tray drying experiment at 1 in. bed height and a drying air temperature of 40°C. Least-squares fit: $\ln(c - c_{\infty}) = 2.541 - 0.069t$. [After Carstensen and Zoglio (1979b). Reproduced with permission of the copyright owner, the *Journal of Pharmaceutical Sciences*.]

The assumptions are frequently realistic, but for shallow beds, i.e., sparingly filled trays, Carstensen and Zoglio (1979b) have found that the slopes in Eq. (IV-4-23) are *not* inversely proportional to a^2. The slopes have an Arrhenius-type dependence on temperature. Equation (IV-4-23) is well adhered to (Fig. IV-15).

It is more likely that in shallow beds, as encountered in tray drying, the limiting rate in the drying is that of the granule itself (which is assumed spherical with radius a). The solution for the diffusion equation is here (Crank, 1970, p. 91) exactly the same as Eq. (IV-4-23), except that the symbol a now denotes the radius of the sphere (in meters).

Example IV-4-2 Carstensen and Zoglio (1979b) reported the following moisture figures† for a granulation in 1.0 and 0.5 in. bed depths at 49°C.

	t (hr)		
	0	1	Equilibrium
at 1 in.	12.6	11.43	0
at 0.5 in.	12.45	11.01	0

Is this drying particle or bed dependent?

† A considerably larger number of points was reported in the article.

Answer At 1 in., $k = \ln(12.6/11.43) = 0.097\ \text{hr}^{-1}$; at 0.5 in., $k = \ln(12.45/11.01) = 0.123\ \text{hr}^{-1}$. Since $k(0.5\ \text{in.})$ is not $4k(1\ \text{in.})$, Eq. (IV-4-23) (with bed height a) does not seem to apply. The two k values are not quite identical but close, and drying is probably dictated by particle drying.

IV-4-8 Fluid Bed Drying

When granules are placed in an air stream and the air velocity is in the correct range, the bed will "fluidize." This is a means of efficient drying, and is accomplished by the system shown in Fig. IV-16. The granulation is placed in a slightly conical container the bottom of which is a mesh screen. The container is placed as shown in the air stream. When the air velocity v is low then the air will simply experience a pressure drop ΔP given by the Carmen–Kozeny equation

$$\Delta P = (q/d^2)\left[\epsilon_b^3/(1-\epsilon_b)^2\right] \tag{IV-4-26}$$

where ϵ_b is the porosity of the bed (the granulation) and d is the diameter. This (Scott *et al.*, 1963, 1964; Mody *et al.*, 1964; Rankell *et al.*, 1964) stays fairly constant as long as v is small.

At a critical velocity of the air v', the implicit fluidization velocity, the bed will fluidize. Increased air velocity will then cause bed expansion (increase in bed porosity). The types of bed structures that are encountered are shown in Fig. IV-17.

At a particular velocity called the entrainment velocity v_e air conveyance will start, i.e., the particles will exit through the stack of the dryer. A plot of porosity versus the logarithm of the air velocity is linear (Fig. IV-18), and it is possible to calculate v_e from such a plot by merely extrapolating to unit porosity. This assumes the drying chamber to be cylindrical. The cross section is A, and if the height of the granulation (of weight S and particle density ρ) is a, then

$$S = A\rho a_0 \tag{IV-4-27}$$

where a_0 is the smallest thickness (height) of the bed (i.e., the height of a hypothetical bed of zero porosity). When the bed porosity is ϵ it is seen that

$$\epsilon = 1 - S/A\rho a \qquad \text{or} \qquad a = S/A\rho(1-\epsilon) \tag{IV-4-28}$$

Note that $\epsilon = 1$ corresponds to $a = \infty$, i.e., entrainment.

In actuality the relation is not quite that simple. Wen and Yu (1966) have shown that (a) ϵ at the fluidization velocity is fairly much the same regardless of particle diameter and (b)

$$\epsilon^{4.7} N_{\text{Ga}} = 18 N_{\text{Re}} + 2.7 N_{\text{Re}}^{1.687}$$

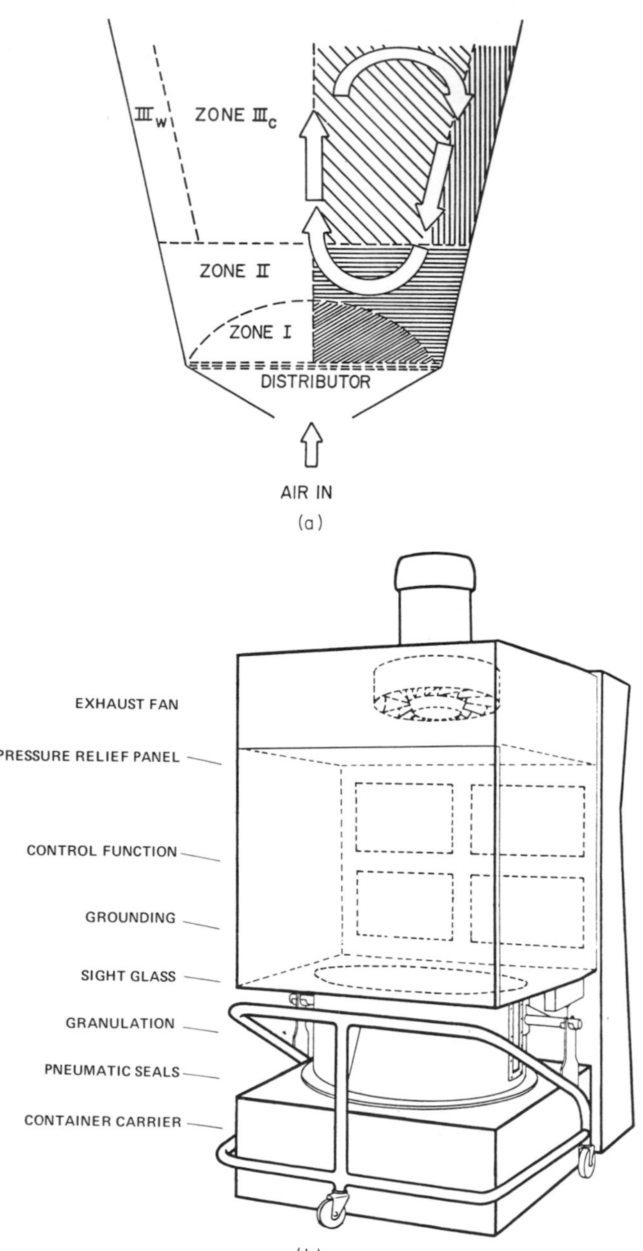

Fig. IV-16 Schematic of a fluid bed dryer: (a) air patterns and (b) equipment used.

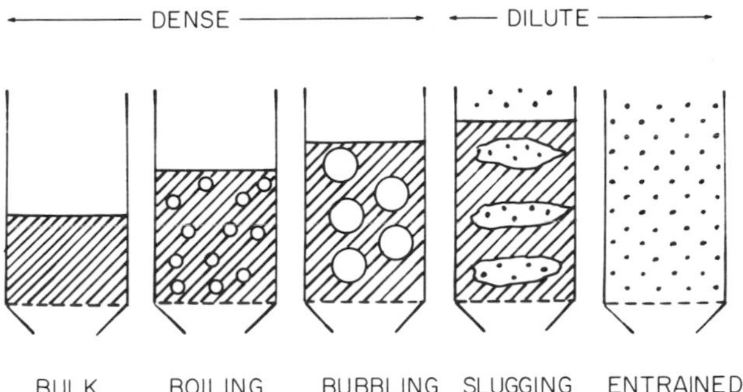

Fig. IV-17 Fluidized bed types: Dense—(a) bulk, (b) boiling, and (c) bubbling; dilute—(d) slugging and (e) entrained.

where the particle Reynolds number $N_{Re} = dv\rho_{air}/\eta_{air}$ and the Galileo number $N_{Ga} = gd^3\rho_{air}(\rho - \rho_{air})/\eta_{air}^2$. Here η is the viscosity. Matsen (1970) has shown that for slug flow the ratio of the maximum height H' at velocity v to the height of the bed at incipient fluidization H_0 is given by

$$H'/H_0 = 1 + (v' - v)/v_B$$

where v_B is the rise velocity of a single bubble.

Leva (1959) has shown that the incipient fluidizing velocity is a function of the particle diameter of the material to be fluidized (Fig. IV-19).

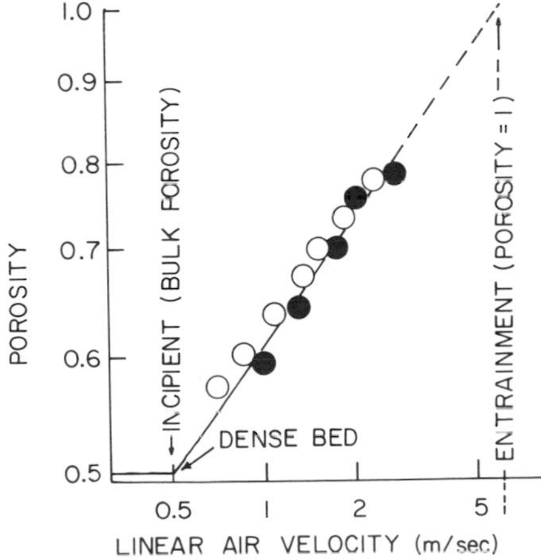

Fig. IV-18 Determination of the entrainment velocity by bed height measurements. Grid diameter: ○, 5 cm; ●, 1.0 m.

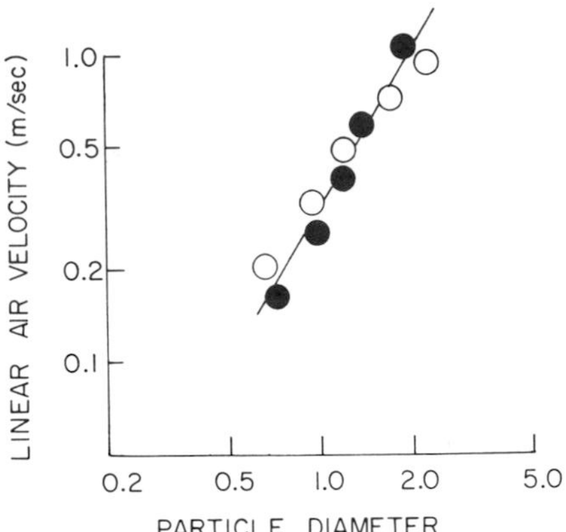

Fig. IV-19 Dependence of the incipient fluidizing velocity on particle size. Grid diameter: ○, 5 cm, ●, 1 m. [After Leva (1959).]

The equation arrived at is given by (Vanacek *et al.*, 1966)

$$v' = 688\, d^{1.82}\left[\rho_{\text{air}}(\rho - \rho_{\text{air}})\right]^{0.94} g^{-1.88} \eta^{-0.88}$$

This equation has been substantiated for rather large particles, (0.25–10 mm diameter), but for particle sizes in the pharmaceutical range it may not necessarily hold either functionally or as far as coefficients and exponents are concerned.

The heat transfer in fluid beds is *very* efficient (of the order of the rate of heat transfer in silver). This is the reason that short drying times are possible ($\frac{1}{2}$ hr). Another convenience is that the drying end point is easily monitored since it occurs when the exit air temperature starts rising. This makes possible automatic shutoff.

Gabor (1966) has shown that the thermal diffusivity α' is a function of the bed diameter D_b, the linear air velocity v, and the incipient fluidization velocity v':

$$\alpha' = 1.255 \times 10^{-4} D_b(v - v')/v'$$

Zoglio *et al.* (1975) have reported on the principles of drying of pharmaceutical granulations in fluid bed driers as a function of time. They demonstrated that the process is rate limited by the diffusion of moisture past the boundary of the granule, which causes the relative humidity RH of the outgoing air less that of the incoming air RH_0 to be log-linear in time (Fig. IV-20):

$$\ln(RH - RH_0) = -\alpha t + \ln(100 - RH_0) \qquad \text{(IV-4-29)}$$

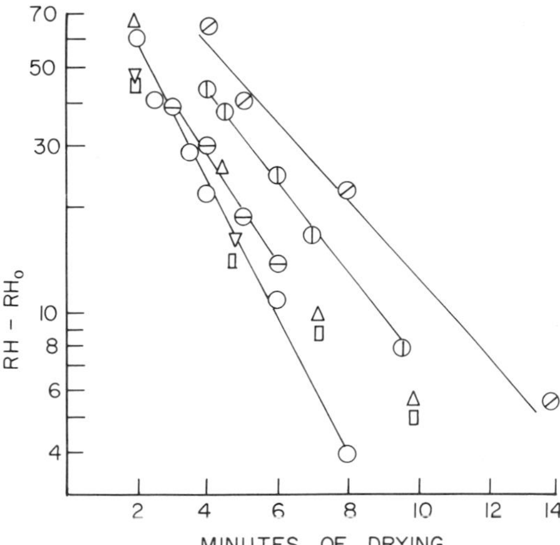

Fig. IV-20 Relative humidity change of air (logarithmic ordinate) as a function of time. Air velocity: ○ 470, ⦶ 460, ⊘ 407, and ⊖ 274 cm/sec. Temperatures: (△) 37°C (460 cm/sec), (▽) 30°C (460 cm/sec), and (□) 25°C (460 cm/sec). [After Zoglio *et al.* (1975). Reproduced with permission of the copyright owner, the *Journal of Pharmaceutical Sciences*.]

The drying rate is a function of granule diameter, as shown in Fig. IV-21.

Particle size distributions are fairly constant during drying. The vivid motion in the fluid bed drying apparently does not translate into large differences in relative velocities of the particles, so that attrition is minimal. The moisture content of the incoming air (more than the temperature) is the important parameter. It has been noted (Vanacek *et al.*, 1966) too, that

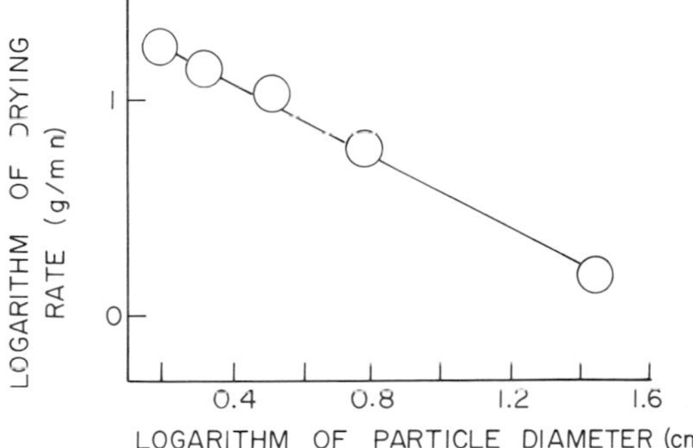

Fig. IV-21 Drying rate in a fluid bed drier as a function of particle diameter. [After Zoglio *et al.* (1975). Reproduced with permission of the copyright owner, the *Journal of Pharmaceutical Sciences*.]

the temperature of the air falls abruptly to the temperature of the bed as the inlet air passes the distributor plate.

It should be noted (as may be seen in Fig. III-10) that fluid bed driers are efficient blenders and that upon discharge it may be assumed that the batch is uniform.

IV-4-9 Fluid Bed Granulation

Fluid bed driers can be modified to become granulators (Mehta *et al.*, 1977; Carstensen and Zoglio, 1979a). In this case a spray nozzle is placed in the center and lower part of the dryer, and the granulating liquid is sprayed in while the air is fluidizing the solid. Hence particles are wetted and dried in the same operation. The granule properties depend, as usual, on the liquid composition or the amount of binder, but here they also depend on air and liquid flow rates. It is often advantageous to use the binder in the solid state (e.g., pregelatinized starch) and granulate with solvent (e.g., water). Mehta *et al.* (1977) found that the particle size distributions are log-normal. The explanation for this is probably the following. The granulation creates aggregates of n particles, where the number n will vary from unity to an upper limit N, which is a function of time t. The aggregates are formed by collision of particle or aggregate with water particles, and the probability of this happening is proportional to the number of particles n, so that

$$dn/dt = bn \tag{IV-4-30}$$

where b is a constant depending on detachment and collision probabilities. The mass M is proportional to n, so that

$$dM/dt = M/g \tag{IV-4-31}$$

where for later convenience the notation $1/g$ [equal to b in Eq. (IV-4-30)] is used. Equation (IV-4-31) can be integrated to give

$$t = h + g\ln(M/\overline{M}) \tag{IV-4-32}$$

where $\overline{M}$ is the mean mass of the aggregates. The first term h is a constant of integration.

The growth time t differs for the different aggregates (Irani and Callis, 1963) and is normally distributed around the average growth time $\bar{t}$, so that a reduced time θ can be defined as

$$\theta = t - \bar{t} \tag{IV-4-33}$$

The fact that θ is normally distributed with mean zero means that

$$f(\theta) = (1/\sqrt{2\pi})\exp(-\theta^2/2) \tag{IV-4-34}$$

Equations (IV-4-32) and (IV-4-33) can be combined to give

$$\theta - \bar{t} = h + g\ln(M/\overline{M}) \tag{IV-4-35}$$

The reduced time is normalized so it is equal to zero when $M = \overline{M}$, in other words

$$\bar{t} = h \tag{IV-4-36}$$

and introducing this into Eq. (IV-4-35) then gives

$$\theta = g\ln(M/\overline{M}) \tag{IV-4-37}$$

Equation (IV-4-37) can now be inserted into Eq. (IV-4-34) to give

$$f\left[g\ln(M/\overline{M})\right] = (1/\sqrt{2\pi})\exp\left[-g^2\ln(M/\overline{M})^2/2\right] \tag{IV-4-38}$$

i.e., the quantity $g(\ln M - \ln \overline{M})$ is distributed normally with mean zero and variance unity, so that M is log-normally distributed, has a mean of $\overline{M}$, and has a standard deviation of $\ln\sigma = 1/g$. $\ln M$ is normally distributed, i.e., $\ln d^3 = 3\ln d$ is normally distributed, so that $\ln d$ is normally distributed. Carstensen and Zoglio (1979a) found that the mean diameter $\bar{d}$ was a function of liquid spray rate R by the relation

$$\bar{d} = a\ln R + b \tag{IV-4-39}$$

where a and b are constants. They found that $\bar{d}$ is not a function of air velocities in the practical range and that the standard deviations are fairly independent of air and liquid flow rates.

Note that as the granulation proceeds the individual particles become heavier and heavier. This means that the fluidization velocity changes, so that it is often necessary to change the air flow rates (in a semicontinuous fashion) as the granulation proceeds.

An attractive feature of spray granulation is that the final milling step is (most often) unnecessary. Hence the total handling is minimized, and the process has gained a fair amount of popularity in the industry. It has been mentioned that fluid bed systems are good blenders. However, for cohesive and electrostatic powders, the deglomeration feature is poor, and may necessitate, e.g., blending or milling prior to fluid bed granulation.

IV-4-10 Vacuum Drying

The potential for drying [as shown for instance in Eqs. (IV-4-18) and (IV-4-19)] is the difference that exists between the vapor pressure P_0 at the wet particle surface or interior and the partial water vapor pressure P_1 in the drying air. The drying in the initial period is also a function of surface area and is inversely proportional to the enthalpy of evaporation L^*. The rate is therefore

$$\text{Rate} = Q(A/L^*)(P_0 - P_1) \qquad \text{(IV-4-40)}$$

where Q is a proportionality factor which is a type of transfer coefficient dependent both on mass and heat transfer. There is, for instance, a dependence on the interfacial energy between air and liquid, γ.

At 25°C water's vapor pressure is about 3.3 kPa. If the air in contact with the granules is reduced in pressure to less than this, then the water on the granulation surface will theoretically boil. (Actually the water will then be present as a saturated solution, with a somewhat smaller vapor pressure.) The bubble formation is a function of γ, and when bubbling starts A will be increased substantially. Rapid drying therefore results by application of vacuum to a drying situation and allows drying at a low temperature (Cooper *et al.*, 1961).

Vacuum drying is *not* a common drying method in the pharmaceutical industry. It has, however, potential for being a good method, and some space will be devoted to it here. A schematic vacuum drying layout (Patterson Kelley Co.) is outlined in Fig. IV-22. It is obviously possible to substitute a V-blender for the double cone blender shown. Note from Fig. IV-23 that wet granulation when placed in the drying blender will, due to the mass effect, cause agglomeration and lumping. This, as shown, can be overcome, by insertion of baffles.

Vacuum drying allows safe drying of products that are heat sensitive. It is obviously more rapid than countercurrent, tray, and truck drying, but is

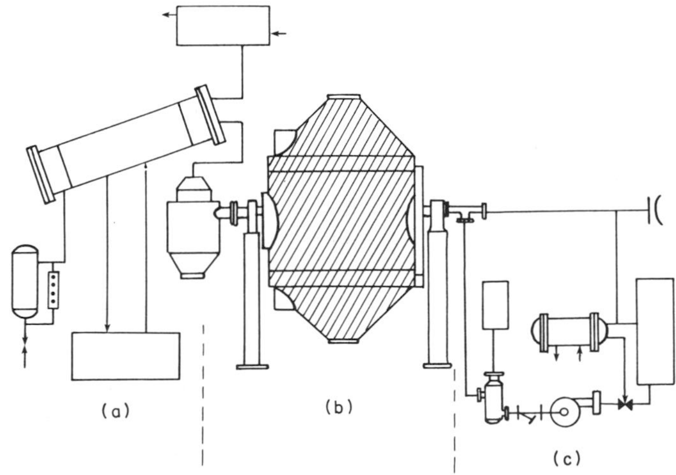

Fig. IV-22 Vacuum drying layout (a) solvent recovery, (b) tumbler vacuum dryer, and (c) jacket heating unit. [Patterson Kelley Company.]

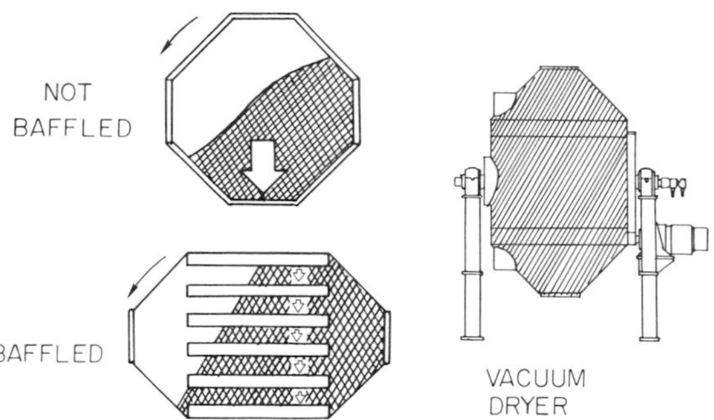

Fig. IV-23 Baffle arrangement for avoiding lump formation during vacuum drying.

not as fast as fluid bed drying. It should be noted that at times (particularly with sugar granulations) a hard crust will form on the outside of the granule, and this will slow down the overall drying, leaving the granule with a moist core. In such cases it may be necessary to discharge, mill and redry. This, of course, is not economic. Often the crust formation can be avoided by reducing the vacuum (increasing the pressure) somewhat. If heat input is used, then altering (lowering) the temperature may also provide a solution.

The temperature sensitivity of a product can be the deciding factor in the selection of a vacuum drying system. There are other advantages. Oxidation, for instance, is minimized. From an environmental point of view, dust entrainment is a serious problem in many drying operations, but it is not so in vacuum drying (at least not to the same degree), because there is a minimum of air movement and a low possibility of particle loss during vacuum drying. This eases compliance with the requirements of OHSA and EPA, and it decreases overall consumption of energy.

IV-5 MILLING

There are three primary types of mills: ball mills, air attrition mills (micronizers), and hammer mills (Fitzmills and micropulverizers).

IV-5-1 Ball Mills and Milling Kinetics

Ball mills are not used extensively in the pharmaceutical industry, but from a theoretical point of view they exemplify the principles of particle diminution. In general, milling is considered a first-order process (Austin *et al.*, 1977; Gardner and Rogers, 1975; Gardner and Austin, 1975; Austin, 1971/1972; Reid, 1965; Jindal and Austin, 1976).

Carstensen *et al.* (1978) and Mehta *et al.* (1977) have found that for pharmaceutical granulations and powders the following treatment holds: When a material is subjected to milling, the mass of material of the original size decreases in time. If a mass w is subjected to milling, then the mass of material having the original diameter d' at a time t is denoted w_a, in such a fashion that

$$\ln w_a = -k't \tag{IV-5-1}$$

In other words the milling treatment is first order (Austin, 1971/1972). There are cases for which this does not hold (Austin *et al.*, 1976) and which are explained by kinetics so as to suggest a process of the type

$$\mathrm{A} \to \mathrm{A}^* \to \mathrm{B} \tag{IV-5-2}$$

Here A* is a particle which has been made softer than the original state but which is still larger in diameter than d'. Equations (IV-5-1) and (IV-5-2) do not seem to apply to pharmaceutical granulations. These although log-linear in the initial phase, frequently give rise to $\ln w_a$ versus time graphs with downward curvature.

The explanation for this may be sought in the following. Kick's law (Parrott, 1970) states that particles of particle diameter d' when milled so as to give rise to powder with a mean diameter d'' require an energy input E of

$$E = C \ln(d'/d'') \tag{IV-5-3}$$

Here C is a proportionality constant relating to mill and substance. At time t there will be a mass w_b of fine diameter d_b, and a mass w_a of the original diameter d_a $(= d')$. The mean diameter of the particles in the sample is therefore

$$d'' = (w_a d_a + w_b d_b)/w \tag{IV-5-4}$$

where $w = w_a + w_b$ is the mass of the sample subjected to milling. The input of energy is proportional to the milling time, so that

$$E = qt \tag{IV-5-5}$$

where q is a proportionality constant. By combining Eqs. (IV-5-3)–(IV-5-5) the following equation is obtained:

$$y = \ln[(w_a/w) + w_b d_b/(w d_a)] = -Kt \tag{IV-5-6}$$

where

$$K = q/C \tag{IV-5-7}$$

The adherence to Eq. (IV-5-6) is shown in Fig. IV-24 (Carstensen *et al.*, 1978). Carstensen *et al.* found that experimentally determined values of d_b correlate well with iterated values, imparting linearity to the data.

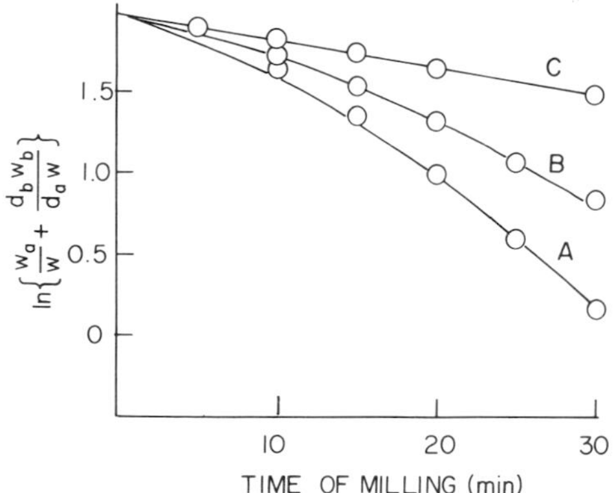

Fig. IV-24 Plotting of milling data according to Eq. (IV-5-6) and with a povidone granulation. Curve A: data as is; curve B: d_b/d_a estimated at 0.25; curve C: d_b/d_a estimated at 0.5. [After Carstensen *et al.* (1978).]

Example IV-5-1 A 10 g powder is ball milled for 10 min. It has an original mesh cut of 20/40 (i.e., is 630 μm in diameter). The fine fraction on milling is 100–200 mesh (110 μm). After 10 min 8 g remains of original size. What is the rate constant K?

Answer Since $w_a = 8$, $w_b = 2$, and $w = 10$, $\ln[0.8 + (2 \times 110)/(10 \times 630)] = \ln 0.835 = -0.18 = -10$ K. Therefore $K = 0.018$ min^{-1}

IV-5-2 Hammer Mills and Particle Size Distributions

In hammer milling, hammers rotate in a milling chamber (a "mill house") into which the powder or granulation is fed. The lower part of the milling chamber is a screen (wire mesh or holed plate) through which the comminuted particles exit.

Steiner *et al.* (1974) have studied the effect of milling on particle size distributions and have found that granules when soft will tend toward normal distributions, but when hard will tend to be log-normally distributed. There are cases in between which follow a Weibull function:

$$\ln(-\ln f) = \lambda \ln(x/k)$$

where f is fraction cumulative over a diameter of x, and where λ and k are Weibull parameters. In cases of *incomplete* granulation or of *overkneaded* granulations there are often two populations of particles, giving rise to bimodal distributions. It is of interest to investigate energy inputs during milling, but direct measurements are difficult (and would require heat loss and amperage measurements, which would be experimentally difficult to balance). An estimate can be made in the following fashion: There is evidence (Martin, 1923; Gross and Zimmerley, 1928) that the area of new surface (ΔA) produced in milling is proportional to the work input E, a concept not in conflict with Kick's law (Eq. IV-5-3) (vonRittinger, 1867; Dallavalle, 1948). This is written

$$E = \gamma \Delta A \tag{IV-5-8}$$

where γ is a proportionality constant.

In the holed plate screen, each individual hole causes a resistance, giving rise to expenditure of energy, and this is set proportional to some negative power of the diameter (d^{-q}). The larger the number n of holes, the smaller the energy expenditure, so that this latter is set inversely proportional to n, with a proportionality constant, β, so that in general

$$E = \beta(d^{-q}/n) \tag{IV-5-9}$$

Combining Eqs. (IV-5-8) and (IV-5-9) and rearranging then gives

$$-q(\ln d) + \ln(\beta/\gamma) = \ln(n\,\Delta A) \tag{IV-5-10}$$

Steiner *et al.* (1974) found Eq. (IV-5-10) to hold well for nine pharmaceutical granulations.

In spite of the variety of distributions encountered after granulation, milling mostly produces log-normal distributions. If it is assumed that there are N particles of initial size x_0 in a sample which is milled, then a certain fraction α is comminuted on each impact (of a hammer mill), rotation (of a ball mill), or collision (in an air attrition mill). For simplicity it will be assumed below that the comminution is a "halving." It may be seen that $N(1-\alpha)$ particles will be left in their original size after one impact and that $N(1-\alpha)^2$ particles will be left after two impacts, etc. After two impacts there will be $4N\alpha(1-\alpha)$ particles of size $x_0/2$ and $4N\alpha^2$ of size $x_0/2^2$. The total number of particles after two impacts is $N(1+\alpha)^2$. The number of particles in the sample after different numbers of impacts (up to m impacts in general) is listed in Table IV-2. It is noted that, then, the number of particles is $N(1+\alpha)^m$ and the possible particle sizes will be from x_0 down to $x_0/2^p$, where p is a number between 1 and m. The number of particles having a size of $x = x_0/2^p$ is seen from the table to be

$$N\binom{m}{p}(2\alpha)^p(1-\alpha)^{m-p} \tag{IV-5-11}$$

The total number of particles is $N(1+\alpha)^m$, so that the fraction having a size of $x = x_0/2^p$ is

$$\Pr(x_0/2^p) = (1+\alpha)^{-m}\binom{m}{p}(2\alpha)^p(1-\alpha)^{m-p} \tag{IV-5-12}$$

The right-hand side of this equation is a binomial distribution which

TABLE IV-2

Number of Particles Remaining after an Indicated Number of Impacts[a]

Size	Impacts[b]				
	0	1	2	3	m
x_0	N	$(N)(1-\alpha)$	$(N)[(1-\alpha)^2]$	$(N)[(1-\alpha)^3]$	$(N)[(1-\alpha)^m]$
$\frac{x_0}{2}$		$(N)(2\alpha)$	$(N)(2)(2\alpha)(1-\alpha)^c$	$(N)(2)(2\alpha)[(1-\alpha)^2]$	$(N)\binom{m}{1}(2\alpha)[(1-\alpha)^{m-1}]$
$\frac{x_0}{4}$			$(N)(4)(\alpha^2)$	$(N)(3)[(2\alpha)^2](1-\alpha)$	$(N)\binom{m}{2}[(2\alpha)^2][(1-\alpha)^{m-2}]$
$\frac{x_0}{8}$				$(N)[(2\alpha)^3]$	$(N)\binom{m}{3}[(2\alpha)^3][(1-\alpha)^{m-3}]$
Total	N	$(N)[(1+\alpha)]$	$(N)[(1+\alpha)^2]$	$(N)[(1+\alpha)^3]$	$(N)[(1+\alpha)^m]$

[a] After Carstensen and Patel (1974). Reproduced with permission of the copyright owner, the Journal of Pharmaceutical Sciences.

[b] Initial number is N.

[c] The general procedure is to note that the amount of size $x_0/2$ disappearing is $N\alpha 2\alpha$; $N2\alpha(1-\alpha)$ remains. The amount of $x_0/2$ produced from size x_0 is twice $N\alpha(1-\alpha)$, so the amount of size $x_0/2$ produced is $(N)(2\alpha)(1-\alpha)$. The total is $(2)(N)[\alpha(1-\alpha)]$.

has been normalized, and this, for large values of m, will approach a normal distribution. The diameters of the particles are, on the other hand, log-linear, as, e.g., $\log 2^{p+1} = \log 2^p + \log 2$ irrespective of the value of p. For large values of m the distribution will therefore approach that of a log-normal distribution.

The argument holds for other situations than halving as long as the particle breaks into a defined equal number of fragments.

IV-6 ANHYDROUS GRANULATION

It is paradoxical (Carstensen and Touré, 1979) that in wet granulation one first adds water and then drives it off. It is obvious that much energy and handling costs could be saved if one could simply blend and compress powders. Several *direct compression excipients* exist, and if the drug content is low, this procedure is feasible. It will be the subject of Section IV-6-2.

IV-6-1 Precompaction

Frequently drug substances are prone to hydrolyze, and must be processed without water. At times a solvent soluble binder can be used, so that a solvent other than water will produce a stable product. Frequently, however, solvents must be strictly omitted in the processing. If the drug content is low, as mentioned, then direct compression is possible. If the drug content is high, and it compresses well, then by mixing it with a suitable excipient it is also possible to directly compress, provided the flow of the powder is good. If this is not the case, then a method denoted precompaction is resorted to. It should be noted that the term precompaction is also used in certain tablet machines for a tamping operation prior to the final compression, and that the term therefore is somewhat ambiguous.

It has been noted, that for a powder to flow well it should have a certain particle size and, preferably, a shape close to spherical. This is accomplished by compressing the formulation in large dies using heavy duty equipment. The compression weights may not be under good control, but the "slugs" are broken up by gentle milling and the portion in a reasonable particle range (e.g., 20/100 mesh) separated. The coarse fraction is remilled and the fine portion recompressed and remilled. In this fashion the powder is converted into agglomerates of a size which will allow flow and hence good weight control of the final tablet. The features *not* accomplished which are imparted by wet granulation are hydrophilization and uniformity. The uniformity of the milled and reblended slug is probably better than that in direct compression but not as good as that obtained by wet granulation.

In chilsonating the same end point is realized as in slugging, but in place of a heavy duty press, a roller compacter (Chilsonator) is used.

IV-6-2 Direct Compression

One might ask what makes a substance autocompressible. Although there is no general answer, one cause is molecular in that the bonding energy in the crystal lattice is usually high in autocompressible substances.

A possibility is alignment. Sodium chloride is autocompressible. It is a cubic crystal and if two surfaces (Fig. IV-25c) (regardless which) are placed side by side and allowed to approach one another, then there is a good possibility (perhaps coupled with a translation) that one face will align so as to match the lattice array of the other. Substances in more sophisticated crystal systems (Fig. IV-25d) have lattice arrangements which differ from face to face, and hence only one out of six (or three) faces will match up with a given face of another crystal.

Surfaces of low rugosity are of importance because there are steps in the crystal face (Fig. IV-25a, b) in the two faces being brought together, and only part of the surfaces will be brought into contact (i.e., approach one another to a distance equal to the lattice constant in the given direction). Hence, on compression, a residual porosity will exist at points

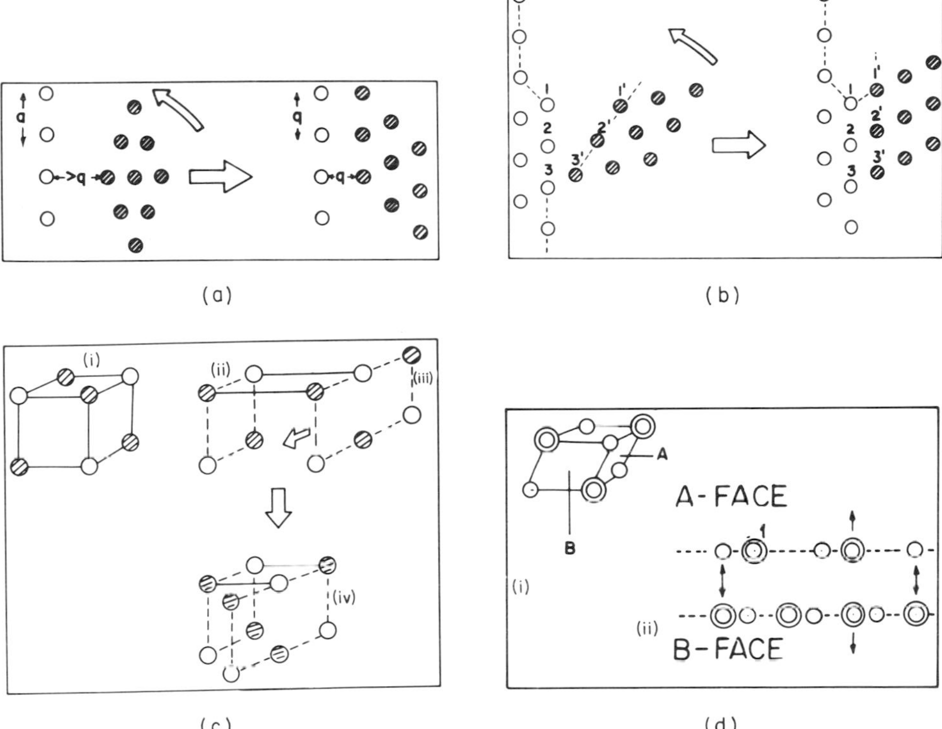

Fig. IV-25 (a) Alignment of two plane surfaces in a crystalline powder. (b) Alignment of two surfaces with a finite rugosity. (c) Alignment by translation of two surfaces in a cubic lattice. (d) Alignment of two nonequivalent surfaces. The arrows indicate attraction *or repulsion*. ○ and ◎ indicate different groups of molecules which are at or close to the surface of the crystal.

of surface irregularities or steps, and through alignment in one part of a bonded surface, there could at another part be "antibonding," so that strains occur (giving rise to cracking on release of pressure).

Direct compression excipients are usually fairly coarse. During the compression, brittle fracture (or plastic deformation) occurs. The creation of new surfaces makes possible bonding between particles. This is particularly effective since the "new" surfaces are noncontaminated and hence are more prone to approach other surfaces at lattice distances.

When a drug substance is noncompressible it is still possible to achieve direct compression if the dose level is low enough. In this case the drug is mixed with a direct compression excipient and simply compressed (after addition of lubricant and disintegrant). The most common direct compression excipients are

(1) lactose (spray-dried or direct compression grade),
(2) anhydrous dibasic calcium phosphate, unmilled, and
(3) microcrystalline cellulose.

If the mixture is primarily excipient, then the compression properties will be those of the excipient, in the opposite case those of the drug. Hence there is a critical concentration of drug beyond which the compressibility is lost. If the drug and excipient are of the same size and loosely stacked in a body centered cubic array, then it is possible to gauge the effect of drug concentration on tablet quality.

In a body centered cubic packing there are six contact points, and the probability of one of these being a drug particle is f if the fraction of drug in the mixture is f. The probability of having two drug particles in contact ("in a row") is therefore $6f$. The probability of having three in a row is $(6f)^2$, etc. If n particles in a row produce a defect then the probability P of a defect is $(6f)^{n-1}$. Values of this for various values of f and n are shown in Table IV-3. The table explains why 15–20% is often the limiting amount of drug which can be incorporated with a direct compression ingredient in a tablet. The argument is simplified, both as far as geometric configuration (body centered cubic) and particle size is concerned, but semiquantitatively accounts for the concentration dependence in direct compression.

The economic advantages (omission of a drying step, simplification, and a great reduction of handling and handling steps) of direct compression are obvious. There are, however, disadvantages. The most severe of these is uniformity, since blending becomes a *very* important parameter and content uniformity is a parameter which needs constant watching in direct compression products. Two other process problems are punch wear and dustiness. Finally, from a product quality point of view, dissolution of hydrophobic substances is usually less rapid from direct compression formulas than from wet granulated products (Finholt, 1974).

TABLE IV-3

Probability of Producing Defects Assuming that n Drug Particles in a Row Produce a Defect

Drug concentration $(100f\%)$	Percent defects			
	$n = 3$	$n = 5$	$n = 10$	$n = 20$
8	23	5	0.1	
10	36	13	1	
13	61	37	11	1
15	81	67	39	14

REFERENCES

Adams, E. F. (1962). Ph.D. thesis, Rensselaer Polytechnic Institute, Troy, New York.

Austin, L. G. (1971/1972). *Powder Technol.* **5**, 1.

Austin, L. G., Trimarchi, T., and Weymont, N. P. (1977). *Powder Technol.* **17**, 109.

Bell, J. R., and Nissan, A. H. (1959). *AIChE J.* **5**, 344.

Bhutani, B. R., and Bhatia, V. N. (1975). *J. Pharm. Sci.* **64**, 135.

Carstensen, J. T. (1973). "Theory of Pharmaceutical Systems," Vol. II, pp. 224–238. Academic Press, New York.

Carstensen, J. T., and Patel, M. R. (1974). *J. Pharm. Sci.* **63**, 1494.

Carstensen, J. T., and Touré, P. (1979). *Labo-Pharma.* No. 285, 201.

Carstensen, J. T., and Zoglio, M. A. (1979a). *Acta. Pharma. Technol. Suppl.* **7**, 17.

Carstensen, J. T., and Zoglio, M. A. (1979b). "Drying." *Am. Pharm. Meeting, Kansas City, Missouri, Nov. 1979.* Abstracts, p. 10.

Carstensen, J. T., Lai, T. Y. F., Flickner, D. W., Huber, H. E., and Zoglio, M. A. (1976). *J. Pharm. Sci.* **65**, 992.

Carstensen, J. T., Puisieux, F., Mehta, A., and Zoglio, M. A. (1978). *Int. J. Pharm.* **1**, 65.

Cooper, M., Swartz, C. J., and Suydam, W. Jr. (1961). *J. Pharm. Sci.* **50**, 67.

Crank, J. (1970). "The Mathematics of Diffusion," 4th ed., pp. 56 and 91. Oxford (Clarendon) Press.

Cruaud, O. (1979). Ph.D. thesis, Université de Paris-Sud (XI), Paris.

Dallavalle, J. M. (1948). "Micromeritics," p. 474. Pitman, New York.

Dingwall, D., and Ismail, S. I. (1977). *J. Pharm. Pharmacol.* **29**, 393.

Duchene, D. (1976). *Labo-Pharma.* **25**, 957.

Finholt, P. (1974). "Dissolution Technology" (L. Leeson and J. Carstensen, eds.), p. 72. Industrial Pharmacy and Technology Section of the Acad. Pharmaceutical Sci., Washington, D.C.

Gabor, J. D. (1966). *Chem. Eng. Prog. Symp. Ser.* **62**, 32.

Gardner, R. P., and Austin, L. G. (1975). *Powder Technol.* **12**, 65.

Gardner, R. P., and Rogers, R. S. (1975). *Powder Technol.* **12**, 247.

Gilliland, E. R. (1938). *Ind. Eng. Chem.* **30**, 506.

Gross, J., and Zimmerley, S. R. (1928). *Am. Inst. Min. Metall. Eng. Tech. Publ.* **46**, 10.

Harwood, C. F., and Pilpel, N. (1968). *J. Pharm. Sci.* **57**, 478.

Healey, J. N. C., Rubinstein, M. H., and Walters, V. (1974). *J. Pharm. Pharmacol.* **26**, 41P.

Irani, R. R., and Callis, C. F. (1963). "Particle Size Measurement, Interpretation, and Application." Wiley, New York.

Jaminet, F., Delattre, L., and Delporte, J. P. (1969). *Pharm. Acta Helv.* **44**, 418.

Jindal, V. K., and Austin, L. G. (1976). *Powder Technol.* **14**, 35.

Jones, T. M., and Pilpel, N. (1965). *J. Pharm. Pharmacol.* **17**, 440.

Jost, W. (1960). "Diffusion," p. 46. Academic Press, New York.

Leva, M. (1959). "Fluidization," McGraw-Hill, New York.

Lewis, W. K. (1921). *Ind. Eng. Chem.* **13**, 427.

Luikov, A. V. (1963). *Int. J. Heat Mass Transfer* **6**, 559.

Martin, B. (1923). *Trans. Br. Ceram. Soc.* **23**, 61.

Matsen, J. M. (1970). *Chem. Eng. Prog. Symp. Ser.* **66**, 47.

Mehta, A., Adams, K., Zoglio, M. A., and Carstensen, J. T. (1977). *J. Pharm. Sci.* **66**, 1462.

Mody, D. S., Scott, M. W., and Lieberman, H. A. (1964). *J. Pharm. Sci.* **53**, 949.

Morgan, R. P., and Yerazunis, S. (1967). *AIChE J.* **13**, 136.

Newitt, D. M., and Conway, J. M. (1958). *Trans. Inst. Chem. Eng.* **36**, 422.

Newman, A. B. (1931). *Trans. Am. Inst. Chem. Eng.* **27**, 203, 310.

Nissan, A. H., George, H. H., and Bell, A. R. (1960). *AIChE J.* **6**, 406.
Oliver, T. R., and Newitt, D. M. (1949). *Trans. Inst. Chem. Eng.* **27**, 1.
Opankunle, W. O., Bhutani, B. R., and Bhatia, V. N. (1975). *J. Pharm. Sci.* **64**, 1023.
Parrott, E. (1970). *In* "The Theory and Practice of Industrial Pharmacy" (L. Lachman, H. A. Lieberman, and J. L. Kanig, eds.), p. 102. Lea and Febiger, Philadelphia.
Pearse, J. F., Oliver, T. R., and Newitt, D. M. (1949). *Trans. Inst. Chem. Eng.* **27**, 9.
Pilpel, N., and Esezobo, S. (1977). *J. Pharm. Pharmacol.* **29**, 389.
Pitkin, C., and Carstensen, J. T. (1973). *J. Pharm. Sci.* **62**, 1215.
Rankell, A. S., Scott, M. W., Lieberman, H. A., Chow, F. S., and Battista, J. V. (1964). *J. Pharm. Sci.* **53**, 320.
Reid, K. J. (1965). *Chem. Eng. Sci.* **20**, 953.
Ridgway, K., and Callow, J. A. B. (1967). *J. Pharm. Pharmacol.* **19**, 1558.
Rubinstein, M. H., and Ridgway, K. (1974). *J. Pharm. Pharmacol.* **26**, 24P.
Rumpf, H. (1958). *Chem. Ing. Tech.* **30**, 144.
Scott, M. W., Lieberman, H. A., Rankell, A. S., Chow, F. S., and Johnston, G. W. (1963). *J. Pharm. Sci.* **52**, 284.
Scott, M. W., Lieberman, H. A., Rankell, A. S., and Battista, J. V. (1964). *J. Pharm. Sci.* **53**, 314.
Selkirk, A. B. (1976). *J. Pharm. Pharmacol.* **28**, 512.
Shafer, E. G., Wollish, F. G., and Engel, C. E. (1956). *J. Am. Pharm. Assoc. Sci. Ed.* **45**, 114.
Shepherd, C., Hadlock, C., and Brewer, R. (1938). *Ind. Eng. Chem.* **30**, 388.
Sherwood, T. K. (1931). *Trans. Am. Inst. Chem. Eng.* **27**, 190.
Sherwood, T. K., and Comings, E. W. (1933). *Ind. Eng. Chem.* **25**, 311.
Steiner, G., Patel, M. R., and Carstensen, J. T. (1974). *J. Pharm. Sci.* **63**, 1395.
Vanecek, V., Markvart, M., and Drbohlav, R. (1966). "Fluidized Bed Drying," p. 28. Chem. Rubber Publ. Co., Cleveland, Ohio.
Veillard, M. (1976). Master's thesis. University of Paris-Sud XI, Faculté de Pharmacie, Paris, France.
vonRittinger, P. R. (1867). "Lehrbuch der Aufbereitungskunde in ihrer neuesten Entwicklung und ausbildung systematisch Dargestellt," p. 82. Ernst and Korn, Berlin, Germany.
Wen, C. Y., and Yu, Y. H. (1966). *Chem. Eng. Prog.* Symp. Ser. **62**, 100.
Zoglio, M. A., Streng, W., and Carstensen, J. T. (1975). *J. Pharm. Sci.* **64**, 1869.
Zoglio, M. A., Huber, H. E., Koehne, G., Chen, P. L., and Carstensen, J. T. (1976). *J. Pharm. Sci.* **65**, 1205.

CHAPTER

V

Tableting and Compression

Compressed tablets are the most common pharmaceutical dosage form. The reasons for this are that (a) they are convenient, compact, easy to carry and ship and (b) they are usually more stable chemically than other dosage forms, since most drugs decompose by hydrolysis and (aside from capsules) other oral dosage forms usually contain substantial amounts of water (or are inconvenient, such as powders for reconstitution).

V-1 INTRODUCTION

Tablet machines are not an invention of recent vintage. They were first introduced in the last century, and they are by now developed to a highly sophisticated level and are high precision tools. This chapter will describe the tableting equipment, the procedures for manufacture and formulation, and basic principles on which formulation is based.

V-2 OPERATIONAL ASPECTS OF TABLET MAKING

Aside from the formulation of tablets, pharmaceutics must deal with their manufacture as well, since the requirements of one frequently dictate the other. For this reason it is worthwhile to conduct an overview of the tablet machines prior to discussing basic solids properties in the compaction process.

V-2-1 Tablet Machines

In the tableting process (Fig. V-1) a granulation (or a powder) is (a) transferred into the cavity or die, (b) retained in the cylindrical cavity by the lower punch, (c) compressed as the upper punch is lowered, and (d) ejected as both punches are raised.

Tablet machines are either excentric (single punch) machines (Fig. V-1) or of the rotary variety (Fig. V-2). In the former, the granulation flows from the so-called shoe into the (single) die in position 1. The feed shoe then moves away, the upper punch lowers to its lowest position and compresses the powder (position 2). Thereafter, both punches move in upward direction (position 3) and force the tablet out. The hopper returns to its original position (and at the same time knocks the tablet, now ejected, into the discharge chute). The level of the granulation in the feed shoe is maintained by means of gravity feed from the bulk in the hopper.

The fill weight (compression weights) of tablets can be adjusted by changing the position of the lower punch. The further down it is, the higher will be the fill weight. This is also a function of the apparent density and, furthermore, depends on the rate with which the powder flows. The compression pressure is adjusted to the desired level by adjusting the position of the lowest point of the upper punch. This, of course, affects the tablet's hardness and porosity.

In rotary tablet machines (so-called rotary presses) there are several dies in a circular arrangement on a die table (Fig. V-2a). The punch heads are guided by and move in cams (Fig. V-2b–d). An evoluted drawing of the die table is shown in Fig. V-2b, which demonstrates how the filling occurs between point A and the scrape-off bar, in other words, at the point where there is contact between granulation (in the feed frame) and the die cavity. The feed frame, in turn, is fed by the powder in the hopper. The granulation is leveled off at the scrape-off bar, and the fill is therefore a function of the position of the lower punch in this position. As the die table turns (from right to left, Fig. V-2b) the die goes past the feed frame and the lower punch is then let drop a bit. (Powder, of course, does not enter into the die past the point of the feed frame.) When the position of the pressure wheels is reached, the lower punch moves upward and the upper punch downward and the tablet is formed. Both punches then rise (guided by the contour of the cam), causing the tablet to eject. The point denoted A′ corresponds to point A (on the back side of the feed frame), which serves to eject (is the ejection bar for) the tablet. Note that the

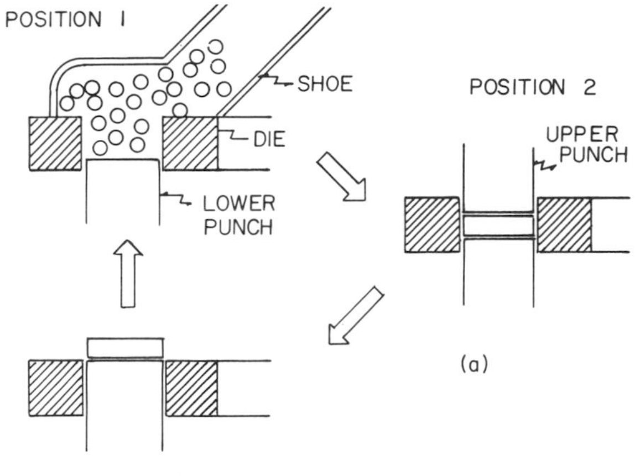

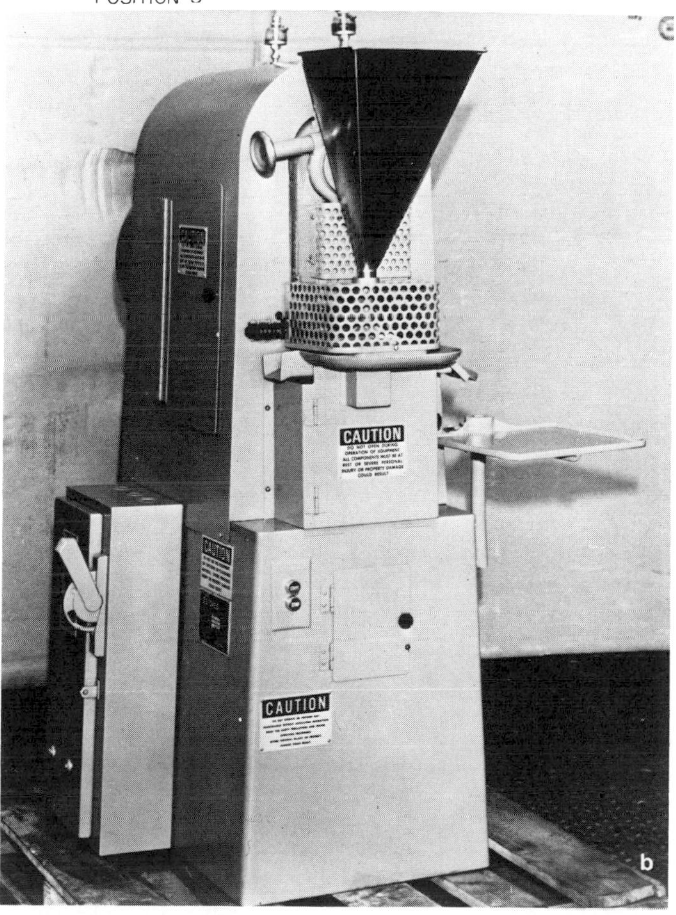

Fig. V-1 (a) Powder flow in a single punch machine (position 1), the compression cycle (position 2), and the ejection step (position 3). (b) Photograph of a single punch machine, the Stokes model 519 (F) press. [Reproduced with the permission of the Pennwalt Corporation, Warminster, Pennsylvania.]

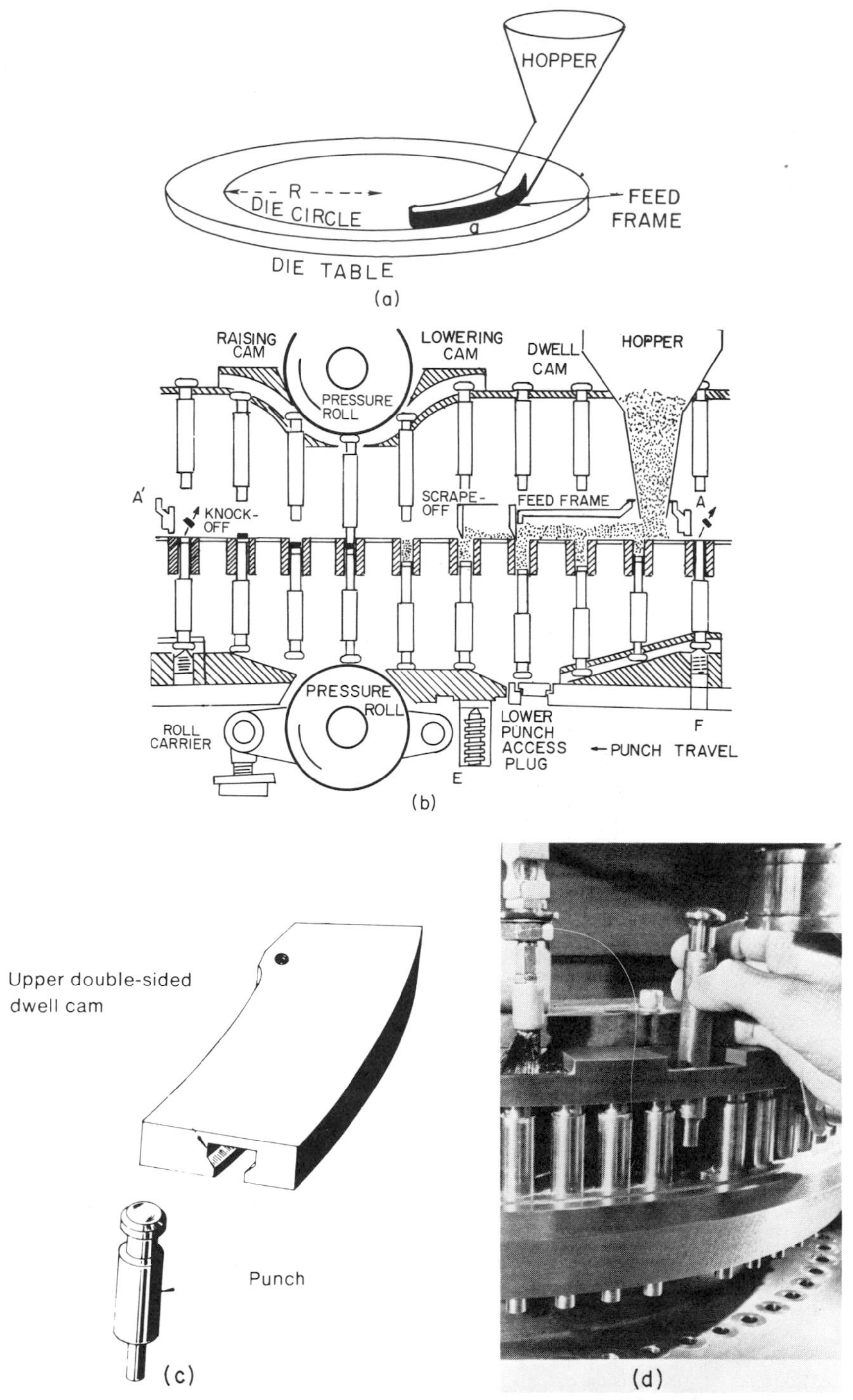

Fig. V-2 (a) Schematic of die table, feed frame, and hopper on a rotary machine. (b) Compression cycle in a rotary machine. (c) Double-sided upper cam used in the Stokes GTP tableting press. (d) Photograph of a punch being installed on a Stokes GTP tableting press. [Reproduced with the permission of the Pennwalt Corporation, Warminster, Pennsylvania.]

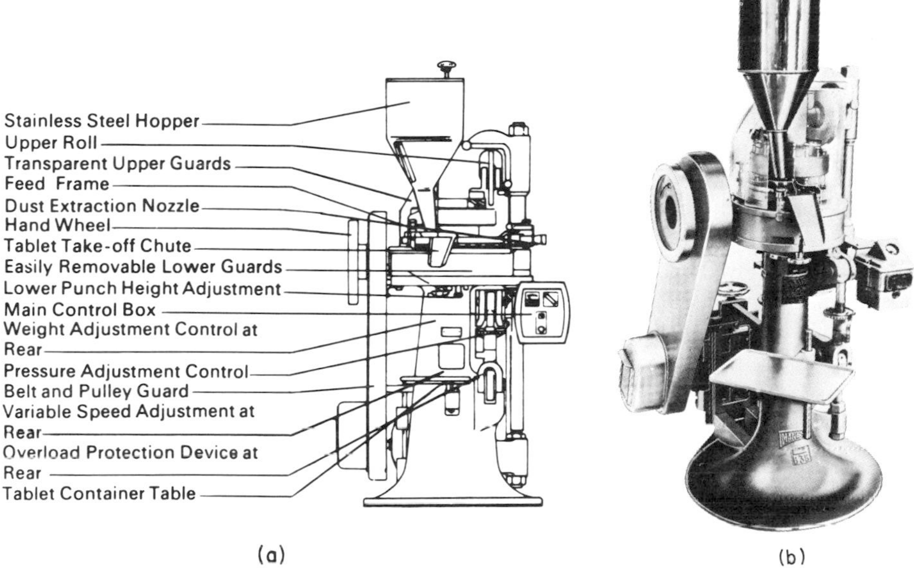

Fig. V-3 (a) Schematic and (b) photograph of a Manesty B3B rotary tablet machine. One model has 16 stations and is capable of producing tablets up to 15.8 mm in diameter. Capacity is 700 tablets per minute. Another model has 23 stations. The filling depth of both machines is 17.4 mm and the highest pressure is 6.5 tons. [Reproduced with the permission of Manesty Machines Ltd., Speke, Liverpool, England, and Thomas Engineering, Hoffman Estates, Illinois.]

weight of the tablet can be adjusted by the screw marked E, the ejection is controlled by screw F (at which point the tablet must be fairly flush with the level of the table), and the compression pressure can be controlled by the relative distance of the pressure wheels.

A single punch machine is shown in Fig. V-1b and a rotary press in Fig. V-3. Some characteristics of these are shown in Tables V-1 and V-2. The production capacities are such that an excentric machine can make about 100 tablets per minute. On the other hand, high-speed presses can produce up to about one million tablets per hour. The principal machines marketed in the United States and Canada are the Stokes† and Manesty‡ machines.

In the simplest rotary presses there is but one hopper and a given number (four or more) of "stations" on the circular die table. In one

† Sharples Stokes division of the Pennwalt Corporation, 955 Mearns Road, Warminster, PA 18974.

‡ Thomas Engineering, Hoffman Estates, IL 60195, and Manesty Machines Ltd., Speke, Liverpool, L24 9LQ England.

TABLE V-1

Outputs, Maximum Pressure, and Number of Stations on Some Presses[a]

Tablet press	Tablets per minute	P_{max} (tons)	Stations
Stokes 519	60–100	4	1
Stokes 525	20–60	20	1
Stokes 511	40–80	1.5	1
Stokes 580	2100		45
	1600		35
Stokes 515	335	10	15
	500	10	16
Stokes DD2	720	15–20	23
Stokes B2	1050	4	16
Stokes 513	4200	4	45
	3260	4	35
Stokes 541	1500–4100	4	41
Stokes 551	1800–5100	4	51
Stokes GTP	3500–10000	10	65
	2800–8000	10	53
	2150–6150	10	41
Stokes Pacer	4200	4	45
	3200	4	35
Stokes 560 Versapress™	2100 (two layer)		45
	1630 (two layer)		35
Stokes 328	1000–2700	10	27
	1200–3300	10	33
	1600–4500	10	45

[a] Reproduced with permission of the Pennwalt Corporation, Warminster, Pennsylvania. Information subject to change without notification.

rotation, therefore, the same number of tablets is produced as there are dies on the machine. The Stokes model 512 rotary press and the model B-2 as well as the Manesty B3B (as shown in Fig. V-3a, b) are examples of this. Somewhat more complicated is the situation in which there are two hoppers (the Stokes model 513-2 rotary press and the Manesty Rotapresses being examples).

Release of entrapped air from tablet granulations and (in particular) tablet powder mixes is of importance, because the release reduces the incidence of capping of lamination of the tablets produced. Machines of the same type as the Stokes 580 Precompression Press and the Manesty Express and Betapress, as well as the high-speed machines of the same type as the Stokes model 610 and the Manesty Mark III, are equipped with precompression features.

Solids to be tableted are divided into three types:

(a) powders that are noncompressible,

TABLE V-2

Outputs, Maximum Pressure and Number of Stations on Some Presses[a]

Tablet press	Tablets per minute	P_{max} (tons)	Stations
Manesty BB3B	760–1520	6.5	27
	924–1848	6.5	33
	1490–2980	6.5	35
	1913–3826	6.5	45
Manesty B3B	350–700	6.5	16
	500–100	6.5	23
Manesty D3B	260–520	10	16
Manesty Betapress	750–1500	6.5	16
	1050–2160	6.5	23
Manesty Express	800–2000	6.5	20
	1200–3000	6.5	30
Manesty Rotapress	888–3552	10	37
	2050–8200	6.5	45
	2500–10,000	6.5	55
	2775–11,110	6.5	61
Manesty RS3	85–224	15	14
	96–256	15	16
	126–336	15	21
Manesty Drycota	176–496 (coated)		
	224–624 (coated)		
Manesty Bicota	320–900 (coated)	8.5	23
	224–624		16

[a] Reproduced with permission of the Thomas Engineering Co., Hoffman Estates, Illinois, and Manesty Machines Ltd., Speke, Liverpool, England. Information subject to change without notification.

(b) powders that are compressible but do not flow well, and
(c) powders that are both compressible and flow well.

For powders of type (a) wet granulation is employed (and this makes them compressible). Powders of type (a) in low dose can also be mixed with direct compression excipients and compressed. Powders of type (b) can be slugged or compacted prior to the actual compression.

It has been mentioned (Section IV-1) that granulation leads to particle enlargement and this in turn leads to better flow. These flow rates are most often higher than for powders that are in category (c). In fact, when such powders are directly compressed the flow is frequently augmented by so-called forced feeders. The Manesty Betapress is equipped with such a system, a two-paddle Rotaflow feeding mechanism, which in turn has a controlled flow inlet. The Stokes series 580 presses have a tunnel feeder (of a pin-type transit level). This guides the granulation or powder directly over the die. It has been mentioned that the flow in the hopper itself can be rate limiting. The Stokes 580 presses have a metering hopper. This lets material, which tends to pack, feed in a uniform fashion. All these steps have as a goal to assure tablet weight uniformity. The Thomas Engineering Sentinel system allows automatic punch adjustment while the machine is operating and adjustment of the compression weights to a tolerance set by a control chart.

A powder of type (b) will form a tablet, but since the flow is not uniform it cannot be direct compressed. It has been mentioned that the flow can be improved by first making a large tablet (a bolus or a slug) on a heavy-duty press (e.g., a Stokes model 516-1, a Manesty RS3, or a Stokes 515-1). The slugs themselves may not have good weight uniformity, but they are milled through a suitable screen, and the fragments (which have a larger particle diameter than the parent powders) have adequate flow to produce tablets of satisfactory weight uniformity.

Much tablet development is done at a product development stage where there are not large amounts of drug available. (This is stage I and II of the clinical trial sequence in drug development.) Scale-up problems into large, high-speed presses can always be anticipated. Government regulations require final production formulas to be very close or identical to the batches used for clinical trials, so that pilot problems are always to be expected in the pharmaceutical industry.

In physical appearance, tablets in general are of a diameter from $\frac{1}{4}$ to 1 in. and are most often cylindrical with a cross section which is circular (Fig. V-4). The tops are either flat (and then most often have a beveled edge) or curved (possessing "crowns") and they can then be either standard concave or deep concave. If a tablet of a cylindrical shape is more

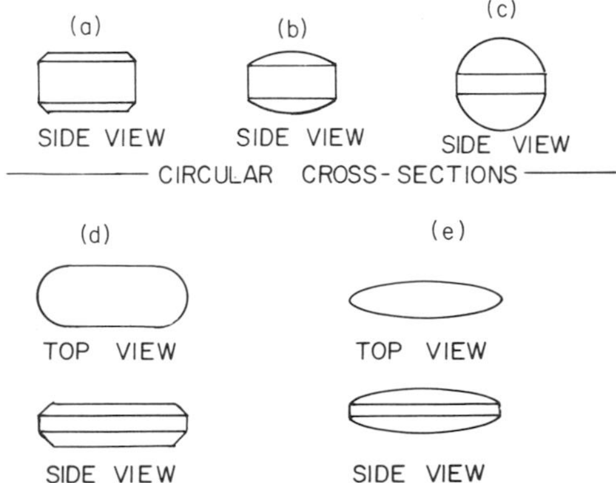

Fig. V-4 Various tablet shapes. [Reproduced with the permission of the Pennwalt Corporation, Warminster, Pennsylvania.]

Fig. V-5 Punches of various tablet shapes. [Reproduced with the permission of the Pennwalt Corporation, Warminster, Pennsylvania.]

than 800 mg, then it is usually difficult for the patient to swallow it, and tablets of this size are usually made in an oblong shape (Fig. V-4d, e and V-5) so that they can be taken "the long way." The tablets having deep concave, oval, and capsule shapes are usually sugar coated. The deep concave (and extra deep concave) shape is specifically suited for coating purposes, because it is almost spherical, and the tablets therefore roll better and interact better with added liquid in a coating pan. Contact areas with other tablets are minimized, thus preventing sticking. Sugar coating makes a tablet easier to swallow but is expensive, and in recent years film coating has often replaced conventional sugar coating.

Oval- and capsule-shaped punches should align properly (as should punches with embossing) and they are therefore keyed or restrained (Fig. V-6). In this manner the punch cannot rotate while on its path along the cam.

Frequently a drug is incompatible with an excipient (Carstensen *et al.*, 1964) or another drug. In the latter case the solids can be separated from one another by several methods. Double- or triple-layer tablets and compression coated tablets (which is a tablet in a tablet) are mechanical means of effecting such a separation. Triple-layer tablets are made in three stages, which requires three hoppers, 120° apart on the rotary die table. Filling is therefore done at three separate times. In the first step, the position of the lower punch controls the fill of the first ingredients (the first layer). Next the bottom limit of the die is the top of the layer of powder which was filled first. In the last stage it is the top of the second layer which is, so to speak, the bottom of the die. It is possible to "tamp" intermediately, and this improves the fill weight precision of the two last layers. The Stokes 580 and the Manesty Bicota are multilayer presses.

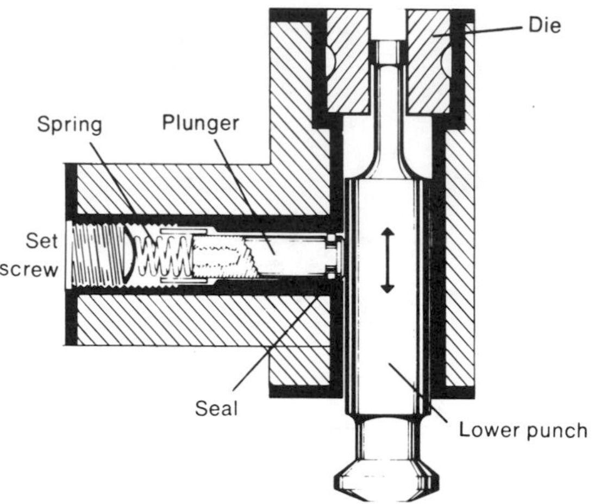

Fig. V-6 Punch restraining system used in the Stokes GTP tableting press. This assures lower punch positioning by use of a spring-loaded plunger. Rotation of the punches is hence prevented, which, for instance, avoids double impression (with embossed punches) and also assures level positioning so that fill weight is better controlled. [Reproduced with the permission of the Pennwalt Corporation, Warminster, Pennsylvania.]

In compression coating (e.g., on a Manesty Drycota) a tablet is first made on the first press (which is one half of the total press assembly). This tablet is transferred over onto a (larger) die in the second half of the machine. This larger die is half filled with an "outer" granulation; then it is filled all the way and the final compression effected. Intergranulation and interlayer bonding, as well as the amount of moisture in the granulations, are important factors in compression coating and multiple-layer compression.

In multiple-layer tablets the precision of fill is not as good as in a conventional tablet. A well defined lower surface (bottom punch) is lacking in the two last fills. Defects are encountered if (a) there is insufficient bonding between layers or (b) the layers are uneven (which can be seen with the naked eye if the layers are of different color). Defects in compression coating are (a) a tablet without a core, (b) a core which is poorly centered and can be detected from the exterior of the tablet, and (c) tablets which have split due to inadequate bonding in the outer layer. The formulation of compression-coated and multiple-layer tablets is difficult.

V-2-2 Tablet Formulation

Before formulating a tablet the external characteristics (size, shape, and thickness) must be known. It is then possible to estimate the tablet weight.

The sum of all the ingredient weights is, of course, the compression weight. It is then possible to estimate the amounts needed of the required ingredients. These are (in general)

drug;
disintegrant, e.g., alginic acid or cornstarch (0–8%);
lubricant, e.g., calcium or magnesium stearate (0–2%);
glidant, e.g., talc (0–1%);
binder, e.g., cornstarch (0–5%);
filler (excipient), e.g., lactose q.s.

A drug is present in a tablet except in the case of a placebo. The amount of the drug, of course, is dictated by its properties. A tablet, after administration to the patient, must *disintegrate*, i.e., fall apart into granules or particles in the gastric juices (or intestinal juices). It will be seen in Section VI-4 that substances which swell when they come in contact with water are used for this. Starch and alginic acid are such substances: These expand enough to break the tablet.

When a tablet is subjected to compression in a die there will result a residual force F which acts against the die wall (Fig. V-7). It is perpendicular to the so-called ejection force E, which is the force exerted by the lower punch to move the tablet out of the die during the ejection phase of the tableting cycle. The two forces are related by

$$E = MF \tag{V-2-1}$$

where M is a frictional coefficient. The main reason for adding a *lubricant* to a tablet formulation is to reduce the magnitude of F. If a powder is not properly lubricated, then the resulting tablets may cap (i.e., the crown will separate, or fall off, from the remainder of the tablet). In milder cases hairline cracks will appear on the wall of the tablets. If very poorly lubricated the tablet may bind up in the die cavity, defying ejection, and the press will stop operating.

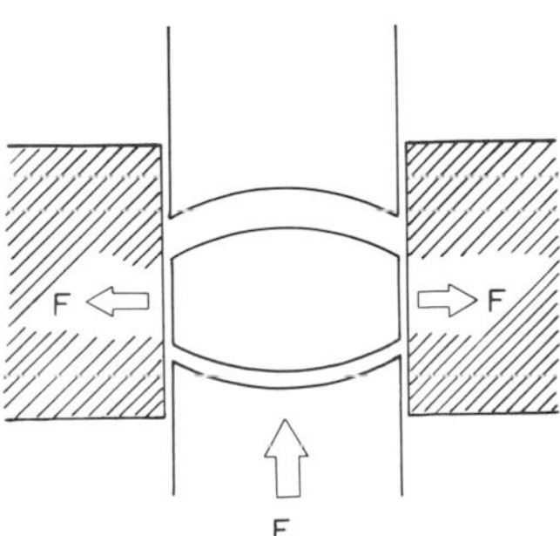

Fig. V-7 Schematic of residual die wall force F and ejection force E.

There are other (formulation) causes for capping also, for instance, if the percentage of fine particles ($< 200\ \mu m$) is too high, then capping is likely to occur. The capper often is formed as the tablet is in the process of emerging out of the die, because at this point it expands. From a practical point of view, the remedies used in overcoming capped tablets when they occur in a batch are variation of lubrication, variation of the machine speed, and reduction of the percentage of fines.

It has been mentioned that a granulation or powder must have good flow properties in order that the produced tablets be of good quality (as judged by variation in weight and uniformity of tablet hardness). *Glidants* may be added to make the flow faster, but in general particle size and surface properties are more important as far as control of flow is concerned. These properties have been discussed in Section II-10. The consistency (uniformity) of the tablet weight is the physical property most affected by poor flow (although tablet hardness uniformity is also affected). The United States Pharmacopeia XVI requires that out of 20 tablets, individually weighed, no more than two may differ from the average by a percentage greater than the following and no deviation can exceed twice this:

tablets weighing 13 mg or less, 15%;
tablets between 13 and 130 mg, 10%;
tablets between 130 and 324 mg, 7.5%; and
tablets more than 324 mg, 5%.

Binders (as will be discussed in Section VI-4) are added to produce granules out of particulate powders, and binders in turn aid in the compression and bonding. As will be described in more detail in Section VI-4, the binders are used in wet granulation (where the binder is added in solution, e.g., as a paste) of type (b) in dry processes, where the powders are blended and then compressed directly, or [type (b)] slugged. The most common pastes are cornstarch (which is used up to 10% and even higher), sucrose, povidone, acacia, and gelatin (5–13% aqueous solution).

The *fillers* used in tablets are usually sugars, inorganic substances, cellulosic derivatives, or sugar alcohols. Lactose, sucrose, dicalcium phosphate, and microcrystalline cellulose are often used as tablet fillers. Most often "excipient" refers to all the nondrug components of the tablet, which includes the fillers.

V-2-3 Factors Affecting Flow of Granulations

Two aspects of the flow rate of powders and granulations are important in tableting: There must be adequate flow from the hopper into the feed frame, and there must be adequate flow from the feed frame into the die.

Powder flow is a function of

(a) particle size,
(b) particle shape,
(c) the roughness of the particle surface,
(d) the chemical nature of the material (e.g., the cohesion), and
(e) the moisture content.

It has been mentioned (Section II-10) that when flow is plotted versus particle diameter a function with a maximum occurs, such as is shown in Fig. V-8 (Carstensen and Chan, 1977). The maximum is denoted (d_m, W_m) and occurs when the diameter is fairly large (400–1000 μm). Therefore, flow problems due to particles being too fine or too cohesive can be solved by particle enlargement. As has already been mentioned, granulation and slugging are the methods usually used.

Ridgway and Rupp (1969) have described the effect of particle shape. The parameter they use for describing the shape is the shape factor: If d is the projected mean diameter, then the surface A and volume V of the particle are $q_1 d^2$ and $q_2 d^3$, respectively. The shape factor is then defined as $G = q_1/q_2$. The shape factor is such that a doubling of it (e.g., from 7.5 to 15) will cause a drop of 20% in the flow rate W.

The Brown and Richards equation (Brown and Richards, 1960) relates the flow rate of a powder to the orifice diameter D:

$$4W/(\pi \rho g)^{0.4} = mD + C \qquad \text{(V-2-2)}$$

Here ρ is particle density, and m and C are constants.

Eq. (V-2-2) shows that the flow rate increase depends on the orifice diameter raised to a power of 2.5. This general type of equation has also been found for dynamic flow (Carstensen and Laughlin, 1979).

The effect of adding fine particles to a powder which is monodisperse has been studied by Danish and Parrott (1971). Their findings are shown

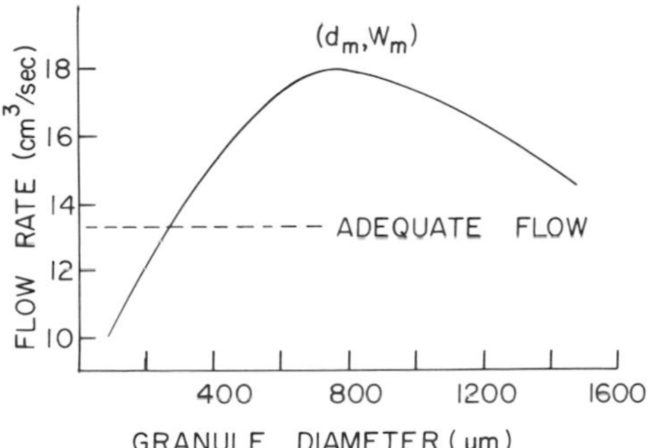

Fig. V-8 Flow rates versus particle diameter. [After Carstensen and Chan (1977). Reproduced with permission of the copyright owner, the *Journal of Pharmaceutical Sciences*.]

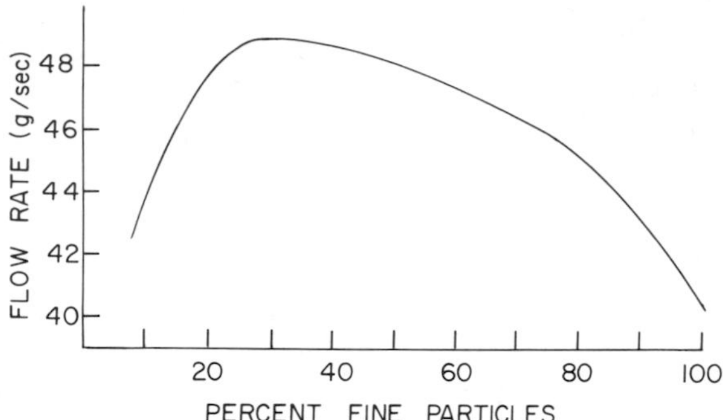

Fig. V-9 Flow rates versus the percentage of fine particles in a binary mixture. [After Danish and Parrott (1971). Reproduced with permission of the copyright owner, the *Journal of Pharmaceutical Sciences*.]

in Fig. V-9. It may be seen that there is a maximum in the curve and that a certain percentage of fines is often advantageous.

V-2-4 Effects of Flow Rate on Tablet Properties

The amount of granulation it is possible to fill into a tablet die is the volume V of the die cavity multiplied by the apparent density ρ' of the powder. If there is a contact time t between the feed frame (which is of length a) and the die, and if the die table is of radius R and rotates at angular speed Ω (rotations per second), then (Fig. V-2a)

$$t = a/\pi 2\Omega R \tag{V-2-3}$$

If the flow rate W is larger than a critical value W', which is given by

$$W' = V\rho'/t = D\Omega 2\pi R/a \tag{V-2-4}$$

then the mass in the die (the fill weight) will be D. If, however, $W < W'$, then the tablet weight will become dependent on flow rate:

$$D = Wa/\Omega 2\pi R \tag{V-2-5}$$

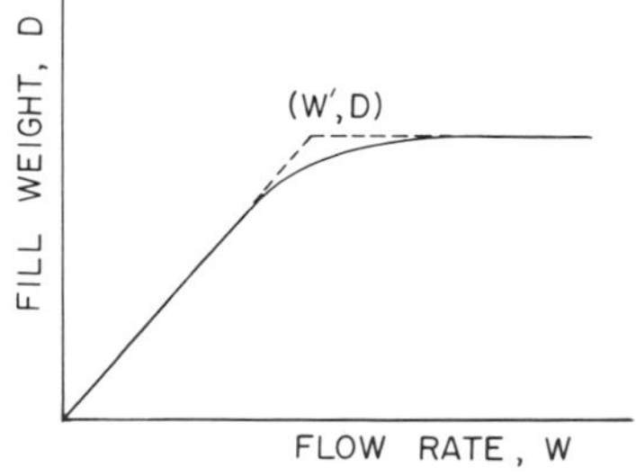

Fig. V-10 Fill weight as a function of flow rate of a granulation or powder.

Equations (V-2-4) and (V-2-5) are shown graphically in Fig. V-10. The sharp break predicted by the equations (the two intersecting dashed lines) does not occur, and on high-speed presses the situation often is that in the transitional region (i.e., the curve below the dashed lines in Fig. V-10).

V-3 PHYSICS OF TABLET COMPRESSION

Marshall (1977) defines the physics of tablet compression as "the compaction of a powdered solid within the confines of a die between two punches by the application of external load." The physics of this process may be simply stated as "the compression and consolidation of a two-phase particulate solid–gas system due to applied forces."

Figure V-11 exemplifies the steps which occur during the formation of a tablet:

(a, b) Rearrangement: As the powder flows into the tablet die it will presumably be of a structure corresponding to the cascaded (untapped) apparent density. The first thing that happens is that the powder rearranges to become a closely packed ensemble, and this is denoted rearrangement, or repacking.

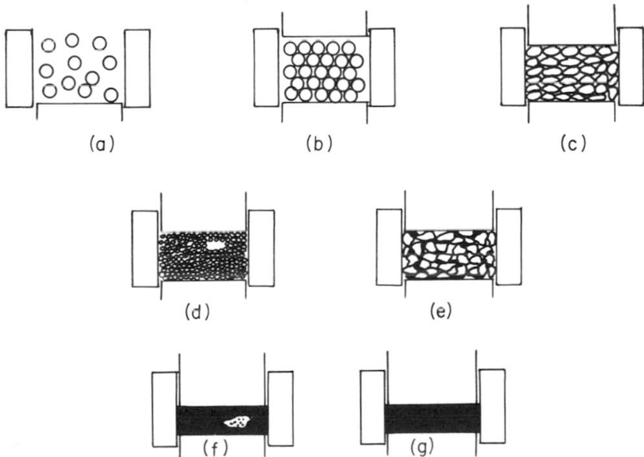

Fig. V-11 (a) Filling of the die in cascaded fashion, (b) rearrangement, (c) elastic deformation, and (d) brittle fracture. It is assumed that the rearrangement step is insufficiently fast, and that a "hole" forms, which eventually leads to the situation in (f). (e) Plastic deformation: The filled-in areas represent "flowing" material. In both (d) and (e) there is substantial void space (air) present. (f) Tablet formed from situation (d). The filled-in space represents a compact with minimum amount of porosity. (g) Properly formed tablet from (e) or from (d) without a hole.

(c) Elastic particle deformation: As the punch comes down on the now most closely-packed arrangement of particles, the particles will start deforming so as to reduce the void space and initially this can be viewed as an elastic particle deformation. In other words, if the punch pressure were released at this particular point, the particles would rebound back into the most closely-packed arrangement.

(d, e) Plastic deformation/brittle fracture: Elastic limits will be reached for the particles and either of two phenomena can occur. Either the particles will undergo plastic deformation or they will fracture.

(f, g) Fusion: The plasticity of the deformed particles (in that particular case) allows for the rearrangement of molecules from one particle with relation to another and for the proper alignment and distance of molecules from various particles so as to form a chemical bond. In the case of brittle fracture new surfaces are formed, which are free of absorbed gas and allow alignment (probably less effectively than in the case of plastic deformation), permitting a pseudo-ideal match of the positions and distances of the molecules in one particle with the molecules in another particle, therefore giving rise to a chemical bond. In both cases the particle contact areas are increased, and this, along with the mobility that the now either fluid or fractured system possesses, is what promotes the bond formation. Some authors have denoted this particular aspect *consolidation*, but this terminology is ambiguous since frequently the repacking or rearrangement mentioned above is also called consolidation.

V-3-1 Rearrangement

The rate of rearrangement of particles is obviously important in the formation of a tablet. If the rate with which a powder rearranges is too slow, then compression can give rise to brittle fracture and plastic flow in certain regions before a close arrangement has been achieved in other regions, as shown in Fig. V-11, and hence when compression is complete a tablet will arise which has a lower local density in that particular area. The manner in which a powder rearranges is shown graphically in Fig. V-12. It has been shown by Takieddin *et al.* (1977) that this follows the equation

$$(V - V_\infty)/(V_0 - V_\infty) = (1 + kn)^{-0.25}$$

In the equation V_0 indicates the initial (cascaded) volume of a powder sample, which is then shaken n times, giving rise to a volume V after n taps. The final volume (corresponding to the most closely arranged packing) is V_∞ and k is a rearrangement constant. Neumann (1967) and Carstensen (1973) have suggested an exponential decay rather than the power function shown above. Once the closest packing is attained further

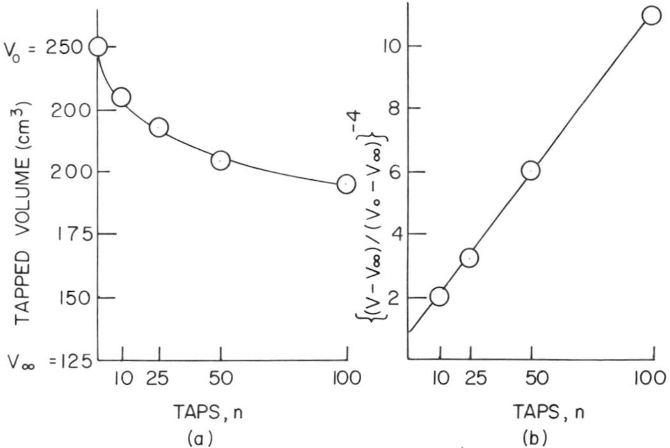

Fig. V-12 (a) Typical rearrangement curve and (b) the data treated according to $(V - V_\infty)/(V_0 - V_\infty) = (1 + kn)^{-1/4}$.

energy input obviously cannot reduce the porosity of the powder bed, and the contact between punch and particles now causes elastic particle deformation.

Example V-3-1 Suppose 250 cm^3 of a powder are cascaded into a graduated cylinder; i.e., $V_0 = 250$ cm^3, and the following results are obtained after the indicated number of taps (n): $n = 10$, $V = 230$ cm^3; $n = 25$, $V = 218.5$ cm^3; $n = 50$, $V = 205$ cm^3; $n = 100$, $V = 193.75$ cm^3. The plateau level is 125 cm^3. Do the data follow a $-\frac{1}{4}$ power law, and if so, what is the rate constant k?

Answer The values of $[(V - V_\infty)/(V_0 - V_\infty)]^{-4}$ are calculated for the four tap values and are shown in Table V-3. These values are plotted

TABLE V-3

Rearrangement Values of 250 cm^3 of Powder

Taps (n)	Volume V (cm^3)	$(V - 125)/(250 - 125)$	$[(V - V_\infty)/(V_0 - V_\infty)]^{-4}$
0	250	1	1
10	230	0.84	2.01
25	218.5	0.748	3.19
50	205	0.64	5.96
100	193.75	0.55	10.93
	125		

versus the number of taps n in Fig. 12b, and it may be seen that a straight line with intercept 1 is obtained, as predicted by the power expression. The slope of the line is 0.1, so that $k = 0.1\ \text{tap}^{-1}$.

V-3-2 Elastic Particle Deformation

Before introducing elastic particle deformation a few definitions may be in order. The terminology used in the following is that of Pollack (1964). Stress is denoted σ and is given by

$$\sigma_1 = \text{stress} = \text{force}/(\text{unit area}) \qquad \text{(V-3-1)}$$

If a solid body is exposed to an external stress it will experience a change in dimension (Fig. V-13) and the strain of the system is defined as

$$\sigma_2 = \text{strain} = (\text{change in dimension})/\text{dimension} \qquad \text{(V-3-2)}$$

Hooke's law states that there is proportionality between the stress and the strain of a system, i.e.,

$$\text{Stress} = E \times \text{strain}\ (l) \qquad (\text{N}/\text{m}^2) \qquad \text{(V-3-3)}$$

where the proportionality constant E is frequently referred to as Young's modulus. In actuality the *modulus of elasticity* is E if the change in dimension divided by dimension is equal to length over length, i.e.,

$$l = \Delta L/L \qquad \text{(V-3-3a)}$$

It is called a *bulk modulus* if it is a change in volume divided by volume, i.e., when

$$l = \Delta V/V.$$

The inverse of the bulk modulus is known as the compressibility κ.

As shown in Fig. V-13, the application of a force to a solid block form can change dimensions depending on the actual restraints of the system. As shown, a force F is applied to a block which is supported on the bottom. The force acts over a surface area A, and the block is deformed in a sense in that it turns into a block with a rhombohedral cross section.

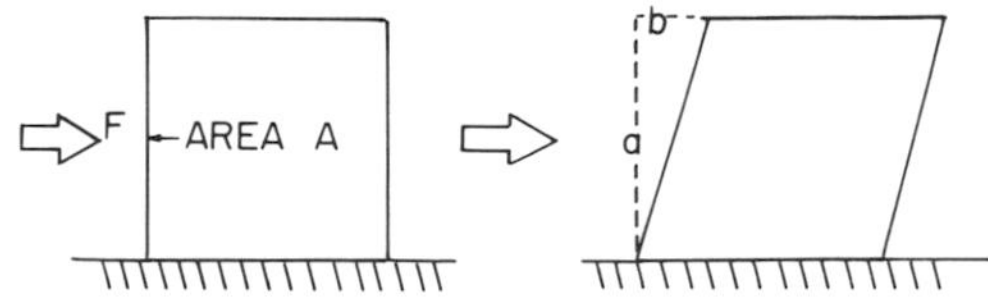

Fig. V-13 Deformation of a cubical block supported below and affected by a force in horizontal direction from the left.

Now the ratio of the change in one direction b to the height a is the result of this shear, and if the change in volume is equal to zero, then

$$G = (F/A)/(b/a) = \tau/(b/a) \quad (\mathrm{N/m^2}) \tag{V-3-4}$$

where $\tau = F/A$ is called the shearing stress.

Example V-3-2 If, for instance, a 0.1 cm^3 crystalline block experiences a 0.01% change in volume upon compression at 10,000 psi, what is the bulk modulus?

Answer Obviously we can write $\Delta V/V = 10^{-4}$, $\sigma_1 = F/A = 10{,}000$ psi. One converts into kg–m–sec units as follows: 1 $\mathrm{lb_{force}} = 0.45\ \mathrm{kg_{force}} = 0.45 \times 9.8\ \mathrm{N} = 4.41\ \mathrm{N}$; 1 in.$^2 = 2.5^2\ \mathrm{cm^2} = 6.25 \times 10^{-4}\ \mathrm{m^2}$; and $\sigma_1 = \mathrm{F/A} = 4.41/(6.25 \times 10^{-4})\ \mathrm{N/m^2} = 7056\ \mathrm{N/m^2}$. Inserting these values into Eq. (V-3-3) then gives $7056 = 10^{-4} E$; i.e., $E = 7 \times 10^7\ \mathrm{N/m^2}$.

Finally, note that the Poisson ratio ν is (Houwink, 1954)

$$\nu = \frac{\text{relative contraction (lateral)}}{\text{relative strain (longitudinal)}} \tag{V-3-5}$$

when the stress is unidirectional. There is a correlation between Young's modulus, the Poisson ratio, compressibility, and the bulk modulus:

$$E = 3(1 - 2\nu)/\kappa = 2G(1 + 2\nu) \tag{V-3-6}$$

The above considerations hold as long as the pressures exerted on the particles are within the *elastic limits*; if, for instance, the compressional force is relieved at any point, the particles will rebound back into their original shape. However, beyond the elastic limits the compressional force will cause the particles either (a) to undergo plastic deformation, which means that if the compressional force is relieved the particles will no longer return to their original shape, that is, the particles attain a pseudofluid state, or (b) to crack, that is, to experience mechanical failure.

V-3-3 Plastic Deformation, Brittle Fracture, and Stress Relaxation

Higuchi *et al.* (1953) have measured the surface area of sulfathiazole compacts available to nitrogen adsorption as a function of the time in the compression cycle (essentially as a function of compaction pressure). They found it first to increase (due to brittle fracture) and then to decrease (due to fusion).

On the other hand, Huffine (1953) simply found the surface of sodium chloride compact to decrease as the pressure was increased. The explanation to these (apparently conflicting) findings is that sulfathiazole bonds in brittle fracture, whereas sodium chloride bonds by plastic deformation.

The effect of stress relaxation has been studied by several authors, notably Shlanta and Milosovich (1964) and David and Augsberger (1977). It is studied by carrying the compression down .to the point where the upper punch is in its lowermost position, i.e., the pressure point, and the punch is then kept there for a certain length of time. Since the compact has been made at a pressure which is increasing as shown in Fig. V-14a, it is not in an equilibrium condition. Hence once the upper punch is kept in that particular position the pressure that will be experienced by the lower punch will vary as a function of time, eventually leveling off when the stress relaxation is complete. Figure V-14 shows the original pressure sensed by the lower punch P_1 and the final pressure P_2. In this region of compression one assumes that there is both elastic deformation and plastic deformation going on, and one can therefore write (the so-called Maxwellian model)

$$l = l_1 + l_2 \tag{V-3-7}$$

Here l_1 is the strain due to plastic flow and l_2 is the strain due to elastic deformation. It can be written directly that for the latter

$$\sigma = El_2 \tag{V-3-8}$$

It is assumed that l_1 adheres to a Newtonian behavior, i.e., the stress is proportional to the strain rate, and the proportionality constant η (sec m^2/kg) is a type of viscosity term. This is written

$$\sigma = \eta\, dl_1/dt \tag{V-3-9}$$

Differentiating Eq. (V-3-7) and introducing Eq. (V-3-8) and Eq. (V-3-9) then yield

$$dl/dt = dl_1/dt + dl_2/dt = \sigma/\eta + (1/E)(d\sigma/dt) \tag{V-3-10}$$

This may be written (when $dl/dt = 0$)

$$d\sigma/\sigma = -(E/\eta)\, dt$$

which integrates to yield

$$\sigma = \sigma_0' \exp[-(E/\eta)t] \tag{V-3-11}$$

or

$$\ln P' = \ln P_0' - (E/\eta)t \tag{V-3-12}$$

Figure V-14 shows data treated in this fashion (Jastrzebski, 1976) and the linearity is obvious, so experiments aimed at monitoring the lower punch pressure after the upper punch is let dwell allow estimation of E, η, and P_0'. Note that σ is assumed to be proportional to P', the pressure sensed by the lower punch.

Hiestand *et al.* (1977) studied die wall and punch pressures in compression followed by decompression followed by recompression, etc. The

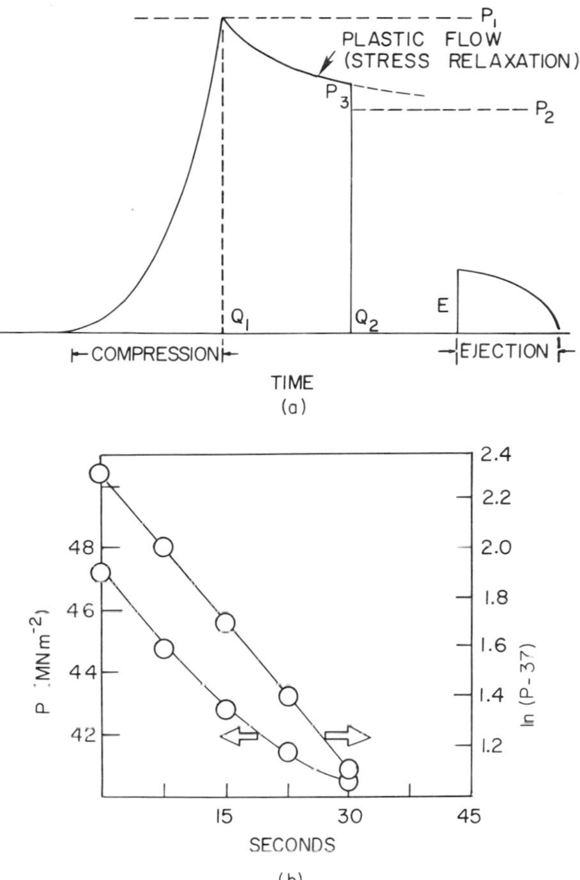

Fig. V-14 (a) Pressure sensed by the lower punch as a function of time during the compression cycle. Here Q_1 is the time when the flight of the upper punch stops, and Q_2 is the time at which withdrawal of the upper punch starts. Stress relaxation is obtainable from the difference between P_1 and P_2 (maximum relaxation) and P_3 (actual relaxation during dwell time). (b) Logarithmic decay curve of the portion from P_1 to P_3 in (a) (see Table V-5).

conclusions drawn were that (a) decompression gives rise to shear deformation in the compact, (b) fracture may or may not occur, and (c) the relief of stress [as also found by Shlanta and Milosovich (1964) and by David and Augsberger (1977)] is a function of time. The results showed that materials that give rise to *slow* release also give rise to capping. They developed a direct test for assessing what they have coined BFP (brittle facture propensity). This is obtained by performing a transverse compression test (as described in Section V-3-5) of a compact with and without a hole in it, the tensile strengths being σ_{T0} and σ_T, respectively. The BFP is

TABLE V-4

Brittle Fracture Propensities[a]

Material	Brittle fracture propensity	Material	Brittle fracture propensity
Methenamine	0.83	Starch, modified	0.27
Erythromycin base	0.65	Lactose (spray dried)	0.18
Ibuprofen	0.40	Microcrystalline cellulose	0.04
Sucrose	0.35		

[a] After Hiestand *et al.* (1977). Reproduced with permission of the copyright owner, *J. Pharm. Sci.*

then defined as

$$\mathrm{BFP} = 0.5\left[(\sigma_{\mathrm{T}}/\sigma_{\mathrm{T0}}) - 1\right] \tag{V-3-13}$$

Note that $0 \leqslant \mathrm{BFP} \leqslant 1$ and that the higher the BFP the more likely is capping of the tablets made from the material. Table V-4 lists BFP values for several materials. One exception to the general trend is acetominophen, which has a low BFP value but a high actual capping propensity. It should be noted, however, that capping is possible within the die or on ejection, and if the latter occurs, then the volume expansion of the compact will be a factor also; in other words, one may consider the propensity for capping as a function of (a) the material (the BFP) and (b) the actual stress on the compact in its formation and ejection. This latter, of course, is material dependent, and if a compact has a low BFP but expands very much, then this expansion may account for the capping.

The Hiestand theory (Hiestand *et al.*, 1977) is exceedingly useful as a classification tool in capping, and future refinements of it will, undoubtedly, pinpoint an exact correlation with observed experience.

Example V-3-3 Data reported by Shlanta and Milosovich (1964) from the stress relaxation of sodium chloride of a particle size finer than 230 mesh, compressed at 18000 psi on a 0.5 in. punch are (approximately), using the nomenclature of Fig. 14a, $P_1 = 47.18\ \mathrm{MN\,m^{-2}}$, after 7.5 sec $P = 44.68\ \mathrm{MN\,m^{-2}}$, after 15 sec $P = 42.75\ \mathrm{MN\,m^{-2}}$, after 22.5 sec $P = 41.37\ \mathrm{MN\,m^{-2}}$, and after 30 sec $P = 40.33\ \mathrm{MN\,m^{-2}}$. Assuming an exponential decay, what is the value of P_2, and would the compact be fully relaxed if ejection took place after 30 sec?

Answer The data are tabulated in Table V-5. Since P_2 linearizes $\ln(P - P_2)$ as a function of time, the correct value is found by iteration. The values 37.0 and 37.3 are shown. Both give fairly good fits and hence P_2 may be assumed to be in this range. The data are shown (using $P_2 = 37$)

TABLE V-5

Stress Relaxation Data for Sodium Chloride Compacts[a]

Time (sec)	P (MN m^{-2})	$\ln(P - P_2)$	
		$P_2 = 37.0$	$P_2 = 37.3$
0	47.18	2.32	2.29
7.5	44.68	2.04	2.00
15	42.75	1.75	1.70
22.5	41.37	1.48	1.40
30	40.33	1.20	1.11

[a] Data from Shlanta and Milosovich (1964).

in Fig. 14b. Since at 30 sec $P = 40.3$ MN m^{-2}, the mass is not fully relaxed at this point.

V-3-4 Fusion

We have already argued in this section that fusion between particles occurs because (a) plastic flow allows the molecules from two particles to become aligned in a configuration (and with distances) closely resembling a lattice (Fig. V-15) and (b) brittle fracture allows realignment of broken particles so that the probability of multiple fractions becoming aligned increases.

Another type of fusion has been suggested, that is, aspcritic bonding. Here melting is presumed to occur at contact points, and on solidification the melt constitutes the bond. Although such bonding probably occurs in

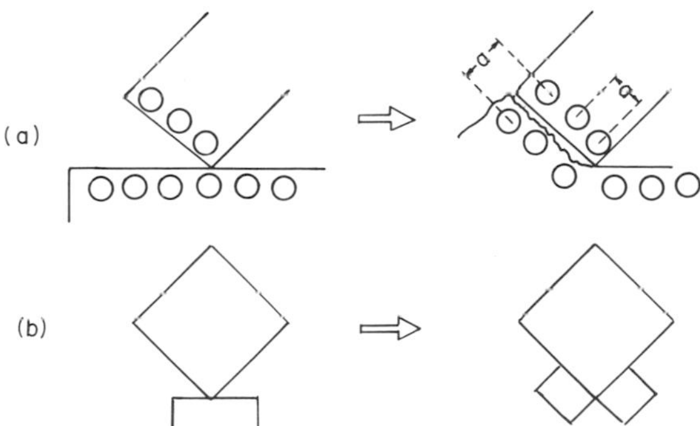

Fig. V-15 (a) Schematic of bond formation in plastic deformation. (b) Schematic of bond formation by brittle fracture.

substances with low melting point (Jayasinghe *et al.*, 1969; Jayasinghe, 1970; York and Pilpel (1972a, b, 1973)), it cannot be a general explanation for bonding in tablets. The melting point (T °K) usually *increases* with increased overall pressure P following the Clausius–Clayperon equation:

$$dT/dP = (V_L - V_S)T_m/\Delta H \qquad \text{(V-3-14)}$$

where V_L and V_S are molar volumes of liquid and solid, respectively, T_m is the melting point at normal pressure (i.e., usually 1 atm), and $\Delta H = H_L - H_S$ is the enthalpy of melting. Since V_L is usually larger than V_S (the density of the liquid is usually less than that of the solid), and since H_L is larger than H_S, it follows that in general dT/dP is positive; that is, the melting temperature increases with increasing pressure. Rankell and Higuchi (1968) and Skotnicky (1953) have shown that when the pressure P is localized to contact points and when the surrounding pressure of the gas phase is 1 atm the melting temperature will always drop on increased pressure and that this *could* be an explanation for bonding. In essence the V_L term in Eq. (V-3-14) is neglected under these circumstances and the equation becomes

$$dT/dP = -V_S T_m/\Delta H \qquad \text{(V-3-14a)}$$

Example V-3-4 A solid compound of molecular weight 200 has $\Delta H = 32$ kJ/mole, a density of 1.5 g/cm^3, and an atmospheric melting point of 120°C. If the compressional pressure is 35 MPa could asperity bonding be expected? The tablet machine temperature levels off at 40°C.

Answer Since $V_S = 1/1.5$ cm^3/g $= 0.7$ cm^3/g $= 0.7 \times 10^{-6}$ m^3/g $= 0.7 \times 200 \times 10^{-6}$ m^3/mole $= 1.4 \times 10^{-4}$ m^3/mole,

$$dT/T = -\left[(1.4 \times 10^{-4})/(32 \times 10^3)\right] dP = -4.4 \times 10^{-9}\, dP$$

This integrates to

$$\ln T = -4.4 \times 10^{-9} P + \alpha$$

The integration constant is calculated from the condition that at $T = 273° + 120°\text{K} = 393°\text{K}$, $P = 1$ atm $= 10^5$ Pa; i.e., $\ln 393 = -4.4 \times 10^{-9} \times 10^5 + \alpha$; so $\alpha = 5.974$. The general equation is

$$\ln T = -4.4 \times 10^{-9} P + 5.974$$

where P is in pascals. When $P = 35 \times 10^6$ Pa, $\ln T = -154 \times 10^{-3} + 5.974 = 5.820$. Hence $T = 337°\text{K} = 64°\text{C}$, which is above the tablet press temperature. It should be noted that the effect of eutectics has not been accounted for in this example.

York and Pilpel have shown that this type of bonding occurs in chloroquine diphosphate, calcium carbonate, and lactose combinations with four fatty acids (York and Pilpel, 1972b, 1973).

As mentioned, asperitic bonding may occur in some instances but is not the generally accepted mode of bonding. While arguments against this type of bonding are commonplace, Hiestand (1978) compared the behavior of the maximum compression stress σ_c applied in forming a compact and the permanent deformation pressure P' in the compact. This was determined by a special impact test (Hiestand *et al.*, 1971). The relative density ρ' (which is the solids fraction of the compact) was also determined.

Plots of $\ln(P'/\rho')$ and also of $\ln(\sigma_c/\rho')$, both versus ρ', are linear with identical slopes; i.e., σ_c and P' both have the same primary cause.

The solid which is in mutual contact (in a cross section) is proportional to the solids fraction ρ'. If the stress concentration factor is constant then the value σ_c/ρ' required to cause melting should also become a constant. However, it was found experimentally that the logarithmic functions given above are linear in ρ' without a plateau, so that asperity melting can be ruled out.

V-3-5 Tablet Hardness

The most important mechanical quality associated with compressed tablets is their hardness, which will be defined shortly. Obviously the term hardness describes the resistance that the compact may offer towards the stresses and strains of transportation and storage. Various means of checking the hardness of the tablets are used routinely in the manufacture of tablets. Most of these consist of placing the tablet the long way between two anvils and applying a pressure on the anvils, in other words, a pressure in the direction of the diameter of the tablet. The tablet hardness may be expressed in either kilograms of force, pounds of force, or some arbitrary unit (for instance, Strong–Cobb units).

If one were to make two tablets of the same formula but with different sizes, other manufacturing conditions the same, it is to be expected that the hardness would differ. Instead of the word "hardness" it is better to employ such terms as fracture force, and in general data can be made uniform (that is, made independent of tablet size) by considering the tensile strength, which for tensile failure equals

$$\sigma_E = 2F/\pi Dt \qquad \text{(V-3-15)}$$

where σ_E is the tensile strength, F is the magnitude of the force applied to the tablet, D is the diameter of the tablet, and t is its thickness. Under point loading, as shown in Fig. V-16, the tensile stress σ_1 is constant over the load diameter (curve i), and hence σ_1/σ_0 is a straight line (σ_0 being the maximum tensile stress). The compressive stress σ_2 (curve ii) and the shearing stress τ (curve iii), however, increase sharply (theoretically ad

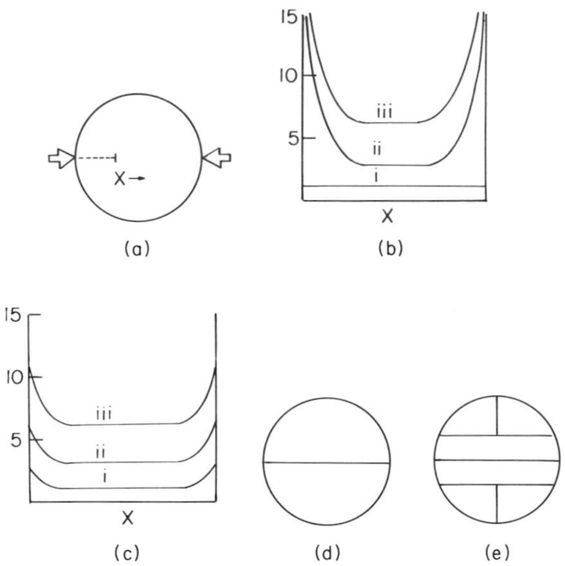

Fig. V-16 (a) Schematic of the diametral test, where X is the distance from one loading point on the diameter to the other loading point. (b) Tension (i) σ_1, compressional stress (ii) σ_2, and shearing stress (iii) τ during point loading. (c) σ_1 (i), σ_2 (ii), and τ (iii) during loading with a platen of finite dimensions. (d) Tensional failure. (e) Primary and secondary tensional failure pattern. [After Hiestand and Peot (1974). Reproduced with permission of the copyright owner, the *Journal of Pharmaceutical Sciences*.]

infinitum) close to the contact point. No loading is a point loading in reality.

In performing the hardness test one observes only failure; i.e., the tablet breaks when σ_1 exceeds σ_{10}, σ_2 exceeds σ_{20}, or τ exceeds τ_0, where the symbols with subscript zero denote tensile strength, compressive strength, and shearing strength, respectively. If (Fig. V-16b) the tensile strength were $1.1\sigma_0$ then the tablet would fail not in a tensile mode but in one of the two other modes.

If the load Q is applied over a width, then the shear and compressive stresses at the loading areas are reduced. If the tensile strength were $0.9\sigma_0$ and the shear strength were $10\sigma_0$ (with an appropriate value for the compressive strength), then the tablet would fail in tensile fashion and the quantity measured would relate to σ_E [Eq. (V-3-15)]. It is easy to observe which type of failure occurs, because, as shown in Fig. V-16d, e, in the case of tensile failure the tablet fails in two halves (Fig. V-16d) or in several clean fractions (Fig. V-16e); otherwise it crumbles into many fragments of different size. Commercial instruments for use in hardness testing have been reviewed by Brook and Marshall (1968) and essentially measure crushing strength in a variety of units, as described above. When the observations are limited to tensile failures the value will distribute itself as a Weibull function (Fell and Newton, 1971). Hiestand and Peot (1974) checked the failure of square compacts. The application of pressure results in some distortion, as shown in Fig. V-17. The distortion (which is related to the magnitude of the compression force) causes molecular displacement, which in turn causes tension in a *horizontal* plane. Here, again, tensile failure causes clean breaks such as those shown in Fig. V-17b–d, whereas shear failure gives the type of split shown in Fig. V-17e. The tensional and compressional stresses hence distribute themselves as shown Fig. V-17f, g. Note that in the square compact the tensile stress is maximum (σ_0) at the *center* of the diameter between the platens. Hiestand points out (as do Fell and Newton, 1971) that the platen width is important. If the width b of the platen is as large as the width a of the compact, i.e., $b/a = 1$, then there will be no development of tension at all ($\sigma_0 = 0$). It has been shown by Berenbaum and Brodie (1959) that maximum shear stresses occur at the edge of the platen, so that (as is also pointed out by Fell and Newton, 1971) with narrow platens, shear strength is exceeded before tensile strength. Padded platens are then recommended and chosen so as to minimize the shear-to-tensile stress ratio τ_0/σ_0. Berenbaum and Brodie found the ratio of τ at the edge of the platen to σ at the center to be a

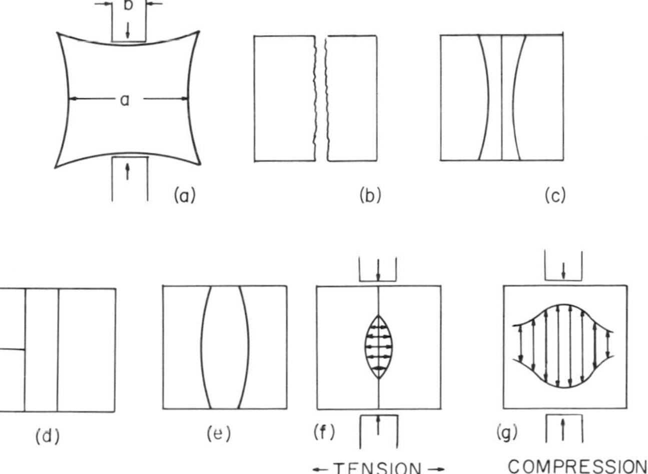

Fig. V-17 Schematic of elastic distortion (oversimplified) caused by transverse compression of a square compact, (b) failure in tension, (c, d) primary and secondary failure in tension, (e) failure in shear, (f) magnitude distribution of tensional stress in compact, and (g) magnitude distribution of compressional stress in compact. [After Hiestand and Peot (1974). Reproduced with permission of the copyright owner, the *Journal of Pharmaceutical Sciences*.]

minimum when $b/a = 0.4$. For this geometry

$$\sigma_0 = 0.16P \tag{V-3-16}$$

where P is applied pressure.

Hiestand *et al.* (1977) checked this relation for square compacts versus the σ_0 values found from diametral tests with $b/D = 0.1$. They found good correlation for some compounds (e.g., sucrose) but poor correlation for others. The results obtained by padding differ from those obtained without padding, and hence values of σ_0 are actually never obtained.

V-3-6 Cappers and Laminated Tablets

Capping and laminating are shown in Fig. V-18. These conditions constitute the most common type of defect in tablet manufacturing. If one asks a "tablet man" what the reasons are for capped or laminated tablets, he will say

(a) that there are too many fines in the granulation (or powder),
(b) that the granulation is too dry, or
(c) that there is not enough lubricant in it.

If he is asked what would be the remedies for correcting the situation he would suggest

(a) slowing down the machine,
(b) polishing the punches, and/or
(c) adding more lubricant.

All these reasons as well as remedies are rational and (in most cases) are the results of many years of experience. This section will attempt to

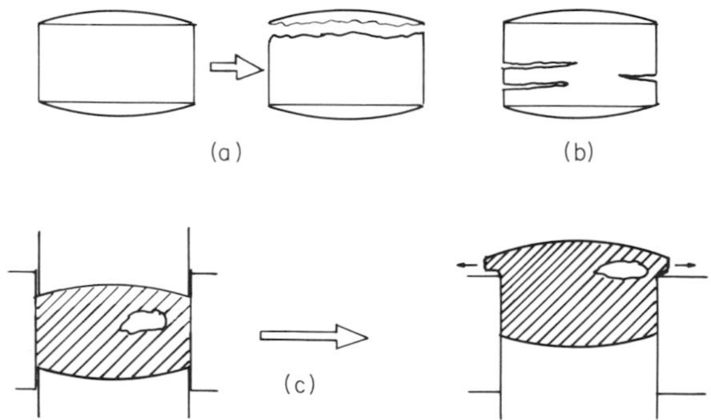

Fig. V-18 (a) Capped tablet. (b) Laminated tablet. (c) Formation of a capped tablet from a low density ("hole") part of a tablet upon ejection.

explain why cappers occur, which will explain why the remedies suggested (sometimes) work.

Schematic representations of capped and laminated tablets are shown in Fig. V-18. The problem of capping is often a scale-up problem. The Pharmacy Research and Development scientists have successfully produced tablets on a single-punch or a low-speed rotary machine. In running this on production equipment at high speeds, there are some factors that are significantly different from what they are on low-speed equipment. One of these is that on a high-speed machine the punch enters the die cavity at a much higher speed than on a single-punch or low-speed rotary machine. Hence the powder flow within the die may not be sufficiently fast and the repacking may be incomplete (Fig. V-18).

Example V-3-5 On a particular single punch machine 90 tablets are produced per minute. One-third of the cycle is taken up by the downward flight of the upper punch. The distance traversed in the flight is 5 cm and the speed of the punch during this period is fairly constant (up to the dwell point). What is this speed?

Answer $60/90 \text{ min} = \frac{2}{3}$ min; i.e., the upper punch travels during a period of $\frac{2}{3}/3 \text{ min} = \frac{2}{9}$ min. The speed is $5/(2/9) = 22.5$ cm/min.

Example V-3-6 On a high-speed machine operating at 100 rpm the downward stroke of the upper punch occurs over $\frac{1}{20}$ of the periphery. The drop is 10 cm. What is the linear punch speed?

Answer One rotation takes 0.01 min and hence the downward punch flight occurs in a time period of $0.01/20 \text{ min} = 5 \times 10^{-4}$ min. The velocity of the packing step hence is $10/(5 \times 10^{-4})$ cm/min $= 2 \times 10^4$ cm/min.

It is seen that there is a *huge* difference between the speeds of the punches and that, therefore, the incomplete packing described in Fig. V-18 may well be the rule rather than the exception in high-speed tableting.

Hiestand *et al.* (1977) have shown that strains magnify about a "hole" and hence a (simplified) situation such as described in Fig. V-18 will give rise to extra strain. Upon release of the upper punch pressure there will be mechanical failure at the hole. Substances which have low values of BFP [brittle fracture propensity, Hiestand (1977)] will not be prone to laminate since they can withstand the strain. Substances with large BFP values, on the other hand, will tend to laminate.

During the extrusion step two factors may convert a laminate to a capped tablet: (a) the friction between compact and die wall is a stress which may cause failure; (b) on the actual exit of the tablet from the die there is (often) an expansion of the tablet, and this stress will cause failure at the hole. In the latter case the crown is the part most likely to fall off.

It should be mentioned that (in particular with embossed punches) the retraction of the upper punch causes an upward pull[†] on the crown and that this is an added stress favoring the dislodgement of the upper crown.

V-3-7 Packing Density and Flow in Compression

Train (1956, 1957) studied the distribution of pressures and densities in magnesium carbonate compacts, and his findings are shown in Fig. V-19. The force was applied by the upper punch in a hydraulic press. Regions of high density and pressure correlate well, as seen in the figure. High values are obtained in the upper corner (i.e., at the edge of the upper punch, where the pressure is applied) and in the center of the tablet mass. Pressure distributions are usually much more uniform when pressure is applied by both lower and upper punch (Fig. V-2b), as is the case in rotary machines, and this is one of the differences in operating characteristics between a rotary and a single-punch machine.

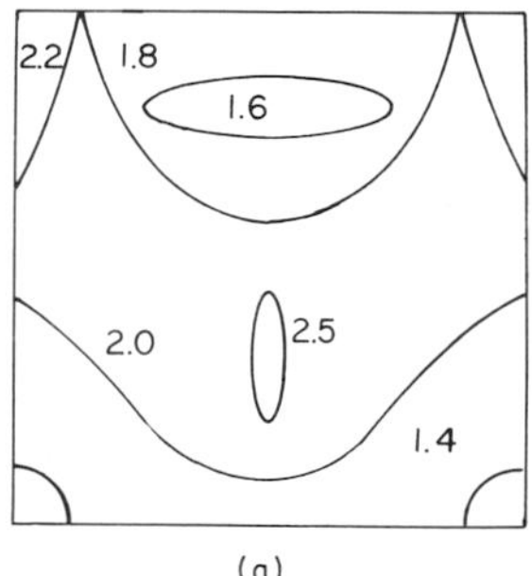

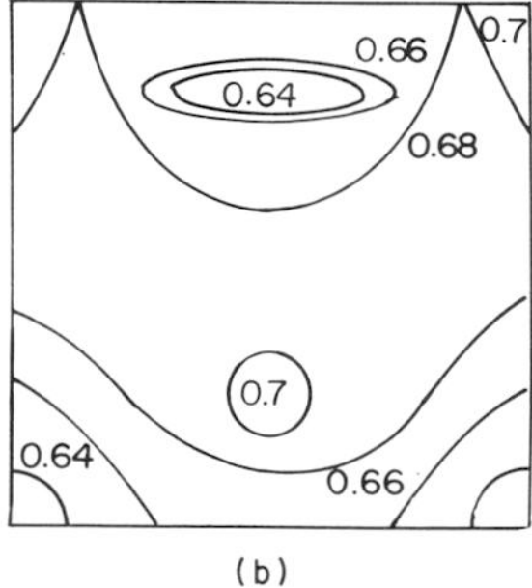

Fig. V-19 (a) Pressure distribution in compacts of magnesium carbonate. The pressures are given in kilograms per square centimeter. (b) Density distribution in the same compacts. Figures are solids fractions. [After Train (1957).]

The pressure P applied in making the compact obviously affects the thickness of the tablet, and this is an extremely important relation in tablet manufacture (although, as will be seen shortly, tablet density is probably a more meaningful parameter than thickness). The thickness h (cm) is generally thought to follow a modification of the Fell–Newton law (Fell and Newton, 1971):

$$\ln[(h - h_\infty)/(h_0 - h_\infty)] = -k(P - P_i) \qquad \text{(V-3-17)}$$

[†] The condition also applies with "picking," where part of the top of the tablet sticks to the upper punch surface. Here, again, polishing the punch surface is the remedial measure.

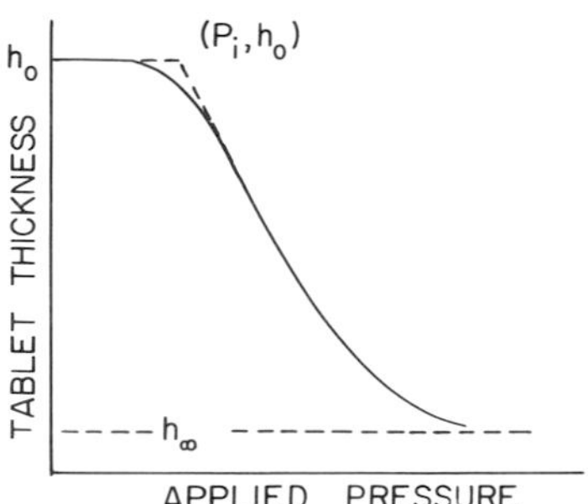

Fig. V-20 Tablet thickness as a function of applied pressure.

This relation is shown (in linear coordinates) in Fig. V-20. One can consider h_∞ as a function of the true density ρ of the tablet in that the (nonporous) mass of the compact is given by

$$D = h_\infty \pi (d/4)^2 \rho \qquad \text{(V-3-18)}$$

where D is the amount of solid in the tablet and d is the diameter. A similar expression relates h_0 to the apparent density ρ'

$$D = h_0 \pi (d/4)^2 \rho' \qquad \text{(V-3-19)}$$

Inspection of Fig. V-20 shows that Eq. (V-3-17) applies beyond $P - P_i$. The reason for this is the sequence in the tablet formation mentioned earlier and shown in Fig. V-11: (a) deposition of a powder bed of cascaded apparent density, (b) then rearrangement to a closest packing (elastic deformation not having been shown in the figure), (c) then plastic deformation or (d) fragmentation, and then (e) fusion. It is assumed that it is the last steps only that apply to Eq. (V-3-17). Hence P_i somehow relates to the elastic limit beyond which the particle no longer returns to the same shape or size when, after deformation, the pressure is released.

The derivation of Eq. (V-3-17) was based on the assumption that the diameter of the tablet is independent of the compaction pressure applied, and this need not be true (as may be concluded as a possibility from Fig. V-18c). Volumes V can be used, however, and in this case the equation becomes

$$\ln[(V - V_\infty)/(V_0 - V_\infty)] = -k(P - P_i) \qquad \text{(V-3-20)}$$

where V_∞ is the true solids volume. Dividing this through by V, and recalling that the porosity ϵ is given by

$$\epsilon = (V - V_\infty)/V$$

Eq. (V-3-20) may be rewritten in the form

$$\epsilon/(\epsilon_0 - \epsilon) = \exp[-k(P - P_i)] \qquad \text{(V-3-21)}$$

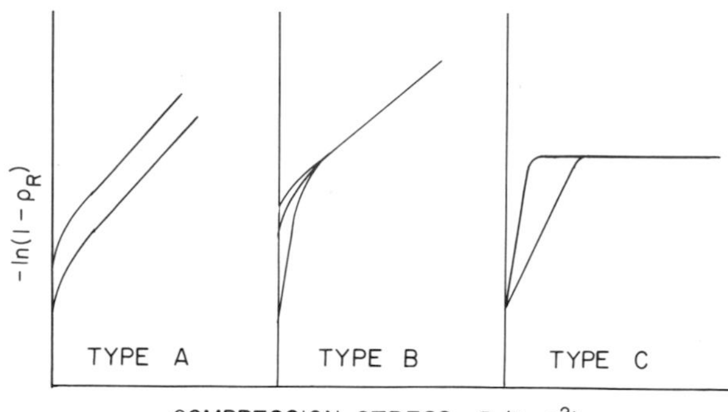

Fig. V-21 The three types of compressional characteristics as described by the Heckel equation. [After York and Pilpel (1973).]

where ϵ_0 is the porosity of the noncompacted mass with volume V_0. This equation is similar in form (but not identical) to the Heckel (1961) equation,

$$\epsilon/(1-\epsilon) = \epsilon_0 \exp(-cP) \tag{V-3-22}$$

Cooper and Eaton (1962) have suggested the formula

$$(V_0 - V)/(V_0 - V_\infty) = c_1 \exp(-k_1/P) + c_2 \exp(-k_2/P) \tag{V-3-23}$$

which, phenomenologically, gives the best fit of all the suggested equations. Both Eqs. (V-3-17) and (V-3-22) are, however, frequently used.

Hersey and Rees (1971) distinguish between two types of behavior in relation to Eq. (V-3-22) and York and Pilpel (1972) add another (type C). These are summarized in Fig. V-21. In all of these the powders are polydisperse. In type A the different mesh cuts have different packing densities throughout; the lines remain parallel. In type B they merge and become identical, and this is thought to be caused by fragmentation during the rearrangement stage. In type C there is no rearrangement during the compression, and densification is solely because of plastic deformation. Lactose by itself will behave as type B, but if fatty acids are admixed will tend toward type C.

Example V-3-7 Treat the date in Fig. V-19 by the Heckel equation [Eq. (V-3-22)].

Answer Table V-6 is constructed correlating (in the first two columns) the pressure P and the packing density ρ_R. The porosity ϵ is $1-\rho_R$, and the quantity $\epsilon/(1-\epsilon) = \epsilon/\rho_R$. The last column then gives the logarithm of this figure. Figure V-22 shows the last versus the first column. Although the correlation is not overwhelming, it is good, particularly when the

TABLE V-6

Data from Fig. V-12 Treated by the Heckel Equation

P (kg/cm^2)	ρ_R	$\epsilon/(1-\epsilon)$		$\ln[\epsilon/(1-\epsilon)]^a$
1.4	0.64	0.36	0.563	−0.575
1.6	0.66	0.34	0.515	−0.663
2.0	0.68	0.33	0.493	−0.708
2.0	0.66	0.34	0.515	−0.663
2.2	0.7	0.3	0.429	−0.847
2.5	0.7	0.3	0.429	−0.847

[a] Least-squares fit $\ln[\epsilon/(1-\epsilon)] = -0.25P - 0.23$ (correlation coefficient, −0.91).

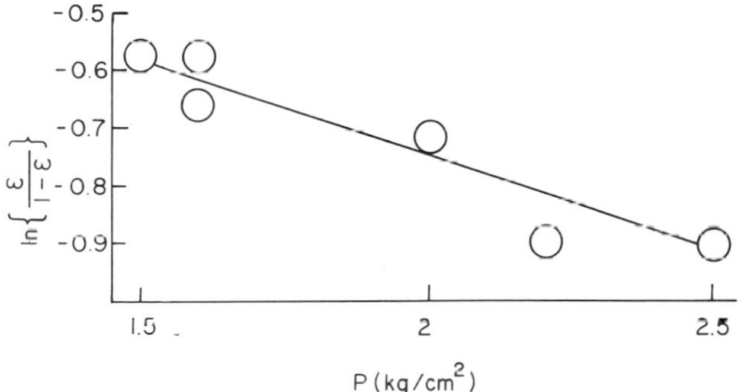

Fig. V-22 Data in Fig. V-19 plotted according to Eq. (V-3-21).

precision range of the data in Fig. V-19 is considered. Note also that the actual local pressure is used as abscissa, not, as dictated by Eq. (V-3-22), the applied pressure.

Shotton and Hersey (1970) have shown that for brittle materials, which fracture (rather than deform plastically), the equation

$$\epsilon\epsilon_0^2 = k_1(\gamma/Pd)^m \tag{V-3-24}$$

holds, which is synonymous with

$$\ln\epsilon = -m\ln P + \ln(k_1\gamma^m/d\epsilon_0^2) \tag{V-3-24a}$$

Here k_1 is a (rate) constant, γ is the surface energy, P is the applied pressure, and d is the particle diameter.

Example V-3-8 A 500 mg direct compression excipient is compressed at four different pressures at a fill weight of 500 mg in a 1/4 in. die. The tablets produced are all of a diameter of 0.25 in. = 0.64 cm. The pressures used (MN m^{-2}) were 2, 4, 6, and 8, and the thicknesses obtained were 0.68,

0.58, 0.53, and 0.52 cm, respectively. The true density of the material is 1.5 and the apparent density is 0.5 g/cm^3. Does the material bond by brittle fracture or by plastic deformation? In the former case what is the value of m?

Answer The void fraction $\epsilon = (V - V_\infty)/V = (h - h_\infty)/h$, so that by calculating h_∞ the porosities can be calculated. Since $h_\infty \pi \times 0.64^2 \times 1.5$ g $= 0.5$ g, $h_\infty = 0.26$ cm. Similarly $h_0 = 0.78$ cm. The porosities are now calculated, and these are tabulated in Table V-7. A straight line of slope is -0.157 is produced when $\ln \epsilon$ is plotted versus $\ln P$. The intercept is -0.377 and the correlation coefficient is -0.993; i.e., there is good correlation, and it may be concluded that the bonding is by brittle fracture.

TABLE V-7

Tabulation of Porosity versus Pressure Data According to Eq. (V-3-24a)

P (MN m^{-2})	h (cm)	ϵ	$\ln \epsilon$[a]	$\ln P$[a]
2	0.68	0.62	− 0.48	0.693
4	0.58	0.55	− 0.60	1.386
6	0.53	0.51	− 0.67	1.792
8	0.52	0.50	− 0.69	2.079

[a] $\ln \epsilon = -0.157 \ln P - 0.377$ ($r = -0.993$).

V-3-8 Compression Cycles

Compression cycles are plots of the radial force (monitored on the die wall) plotted as a function of the actual force (compression force). A fair amount of literature has been published on this subject (Long, 1960; Schwartz and Weinstein, 1965; Perelman and Roman, 1971; Schwartz and Holland, 1969; Koerner and McCabe, 1972; Paris *et al.*, 1975; Leigh *et al.*, 1967; Strijbos *et al.*, 1977).

The Poisson ratio ν has been defined previously [Eq. (V-3-5)]. In a compact

$$\nu = (\text{lateral strain})/(\text{longitudinal strain})$$

In a tablet (Fig. V-7) an applied (longitudinal) pressure P causes a (lateral) die wall pressure F and a pressure P_1 transmitted to the lower punch. Within the elastic limit the strain is proportional to the stress (Hooke's law), so $\nu = F/P$ or

$$F = \nu P \tag{V-3-25}$$

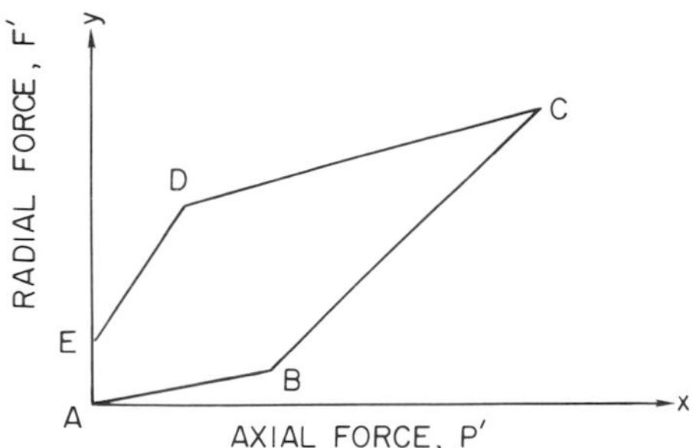

Fig. V-23 Radial and axial recorded pressure cycle.

In instrumented presses one may monitor the die wall and applied pressures as functions of time and the initial curve (AB in Fig. V-23) adheres to Eq. (V-3-25) with some modification. It should be noted that softer materials have larger values of ν. With harder materials ν will approach 0.3. For low ν values one would therefore expect bonding by brittle fracture and for high values bonding by plastic deformation.

In actuality lines AB never go through the origin (Marshall, 1977) but intersect the x axis at a small positive value. The slope is also different from that predicted by Eq. (V-3-25), and the curve AB is more realistically represented by

$$F' = [\nu/(1-\nu)]P' \qquad \text{(V-3-26)}$$

where F' and P' denote the radial and axial forces, respectively.† If the elastic limit is high, then on the removal of the load the original shape(s) will (almost) be regained. Some bonding can of course occur during the elastic deformation, but unless this bonding is very strong the rebound will result in structural failure.

If forces are applied beyond point B (which represents the elastic limit), then, as mentioned, brittle fracture or plastic deformation will occur. If plastic deformation occurs, then the powder behaves like a liquid, so that $F' = P'$. If the yield value is S, then the equation for line BC becomes

$$F' = P' - S \qquad \text{(V-3-27)}$$

If, on the other hand, brittle fracture occurs, then the mass will behave like a Mohr body. Recall from Section II-8 that at failure (on a powder yield locus) the shearing stress τ is (approximately) linear in the normal

† Forces will be denoted by primes and pressures (F and P) without primes, except that P^* denotes maximum force and P_{max} maximum pressure.

stress σ and equals

$$\tau_n = C + \mu\sigma_n \tag{V-3-28}$$

where C is cohesion and μ is a frictional coefficient. To develop the relationships between F' and P', the following notation will be used:

$$N = \nu/(1-\nu) \tag{V-3-29}$$

$$M = (1-\mu)/(1+\mu) \tag{V-3-30}$$

$$Q = 2C/(1+\mu) \tag{V-3-31}$$

$$K = 2C/(1-\mu) \tag{V-3-32}$$

If a stress is increasing and exceeds C [Eq. (V-3-28)], then failure occurs. At failure τ_n (in the plane of shear) is equal to $(\sigma-\tau)/2$ and the normal stress $\sigma_n = (\sigma+\tau)/2$. Inserting these values in Eq. (V.3-28) gives

$$(\sigma-\tau)/2 = C + \mu(\sigma+\tau)/2 \tag{V-3-33}$$

which may be written

$$\tau = [(1-\mu)/(1+\mu)]\sigma - 2C/(1+\mu) \tag{V-3-34}$$

It is conventional at this point to assume that all stresses are proportional to applied and observed forces by *the same* proportionality constant, and if so, one may write Eq. (V.3-34) as

Line BC:

$$F' = MP'' - Q \tag{V-3-35}$$

Point B of Fig. V-23 is the intersection between AB and BC and hence its abscissa x'_B is given by: $Nx'_B = Mx'_B - Q$, i.e.,

$$x'_B = Q/(M-N) \tag{V-3-36}$$

Point C is associated with the maximum pressure P^* and has the coordinates:

Point C:

$$(P^{*\prime}, F^{*\prime}) = (P^{*\prime}, MP^{*\prime} - Q) \tag{V-3-37}$$

The radial force decay (Leigh *et al.*, 1967) after pressure is released at C follows the line CD, which is parallel to AB (i.e., has slope N) and goes through C $(P^*, MP^* - Q)$, i.e., its equation is

Line CD:

$$F' = N(P' - P^*) + (MP^* - Q) \tag{V-3-38}$$

Introducing the quantity

$$U = (M-N)P^* - Q \tag{V-3-39}$$

this may be written

Line CD:

$$F' = NP' + U \tag{V-3-40}$$

At point D the curve will change and have the equation

Line DE:

$$F' = (1/M)P' + K \tag{V-3-41}$$

where $K = 2C/(1-\mu)$ is the residual die wall force.

Since D is the intercept between ED and DC, its abscissa x_D is given by $(1/M)x_D + K = Nx_D + U$, i.e.,

$$x_D = (U - K)/\left[(1/M) - N\right] = VP^* + Z \tag{V-3-42}$$

where V is given by

$$V = (M^2 - MN)/(1 - MN) \tag{V-3-43}$$

and Z by

$$Z = 4CM\mu/(1-\mu^2)(1 - MN) \tag{V-3-44}$$

The area of the hysteresis loop is

$$\begin{aligned} S &= \int_0^{x_D} \left[(x/M) + K\right] dx + \int_{x_D}^{P^*} (Nx + U)\, dx \\ &\quad - \int_0^{x_B} Nx\, dx - \int_{x_B}^{P^*} (Mx - Q)\, dx \\ &= P^{*2}\left[\tfrac{1}{2}(M - N)\left(1 - \frac{M \quad N}{(1/M) - N}\right)\right] \\ &\quad + P^*\left[(K + Q + (1/M) - NZ)V - Z(M - N)\right] \\ &\quad + \left[(1/2M)Z^2 + KZ - \tfrac{1}{2}NZ^2 + ZQ + \tfrac{1}{2}x_B^2(M - N) - Qx_B\right] \end{aligned} \tag{V-3-45}$$

In this case the area, therefore, is a function of P^{*2}.

The second case which may be visualized (when P' exceeds x_B) is where the yield stress of the solid in shear is constant, i.e., is not dependent on how large the principal stresses are. Long (1960) and Leigh *et al.* (1967) follow a treatment whereby the following holds (to distinguish this case the symbols y and x are substituted for F' and P'):

Line AB:

$$y = Nx \tag{V-3-26$'$}$$

Line BC:

$$y = x - C \tag{V-3-46}$$

Point C:

$$(P^*, P^* - C) \tag{V-3-47}$$

Point B: This is the intersection between lines AB and BC, so that x_B is given by $Nx_B = x_B - C$; i.e.,

$$x_B = C/(1 - N) \tag{V-3-48}$$

Line CD: This has a slope of N and goes through $(P^*, P^* - C)$; i.e.,

$$y = Nx + (1 - N)P^* - C \tag{V-3-49}$$

Point D: At point D, "the radial force is greater than the axial force by an amount equal to" C (Leigh *et al.*, 1967); i.e., at point D

$$y_D = x_D + C = Nx_D + (1 - N)P^* - C$$

or

$$x_D = P^* - 2C/(1 - N) \tag{V-3-50}$$

Hence when $P^* > 2C/(1 - N)$ there exists a point D, since x_D in this case has a positive value.

Line DE: This will pass through $(P^* - 2C/(1 - N),\ P^* + C - 2C/(1 - N))$ and have unit slope; i.e.,

$$y = x - P^* - [2C/(1 - N)] + P^* + C - 2C/(1 - N) = x + C \tag{V-3-51}$$

Note that the residual die wall pressure will be C.

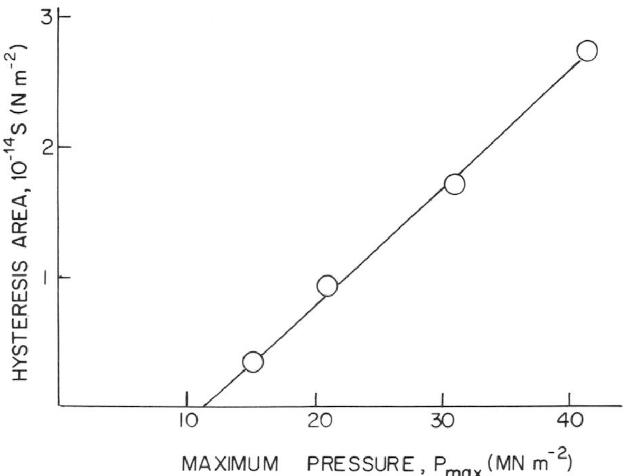

Fig. V-24 Area under the hysterisis loop of sodium chloride compressed at different maximum pressures P_{max}. [Data from Leigh *et al.* (1967, Fig. 3), slightly modified. Reproduced with permission of the copyright owner, the *Journal of Pharmaceutical Sciences*.]

So in the second case, the area of the hysteresis loop becomes

$$S = \int_0^{x_D} (x + C)\,dx + \int_{x_D}^{P^*} [Nx + (1 - N)P^* - C]\,dx$$

$$- \int_0^{x_B} Nx\,dx - \int_{x_B}^{P^*} (x - C)\,dx$$

$$= 2CP^* - C^2[2.5/(1 - N)] \tag{V-3-52}$$

In this case, the area in the hysteresis loop is proportional to the maximum force P^* or maximum pressure P_{max} employed. An example of this is shown in Fig. V-24. Since the slope is $2C$ it is possible to calculate C from the slope. The intercept is $-2.5C^2/(1 - N)$, so that N can now be calculated.

V-4 LUBRICATION AND FORCE TRANSMISSION

During one-sided compression the compression force P' is conveyed by the downward flight of the upper punch. The force sensed by the lower punch, P_1', is less than P'. In the powder there will be an axial force F' which acts on the wall of the die. There will be a frictional force F_f' between the wall of the die and the solid under compression. A force balance in vertical direction then gives (Shotton and Hersey, 1970)

$$F' = P' - P_1' \tag{V-4-1}$$

Application of Eq. (V-2-1) then gives

$$F_f' = \mu F' \tag{V-4-2}$$

In the following L denotes the depth of the solid in the die, D denotes the die diameter, and η denotes the radial stress σ_2 divided by the axial stress σ_1. The general situation about an infinitesimal length dx is shown in Fig. V-25. The force on the die wall is the stress σ_2 times the area i.e., $F' = \sigma_2 \pi D\,dx = \nu\sigma_1 \pi D\,dx$ (since $\sigma_2/\sigma_1 = \nu$). The vertical force is $P_x' - P_{x+dx}'$ which equals $-d\sigma_1$ times the area, i.e., $-(D^2\pi/4)\,d\sigma_1$. Making use now of Eq. (V-2-2), one gets

$$\mu\nu\sigma_1 \pi D\,dx = -(D^2\pi/4)\,d\sigma_1 \tag{V-4-3}$$

This may be written

$$d\sigma_1/\sigma_1 = -(\mu\nu 4/D)\,dx \tag{V-4-4}$$

Integrating this equation from 0 to L then gives

$$\ln\sigma = -(\mu\nu 4/D)L + \ln\sigma_0 \tag{V-4-5}$$

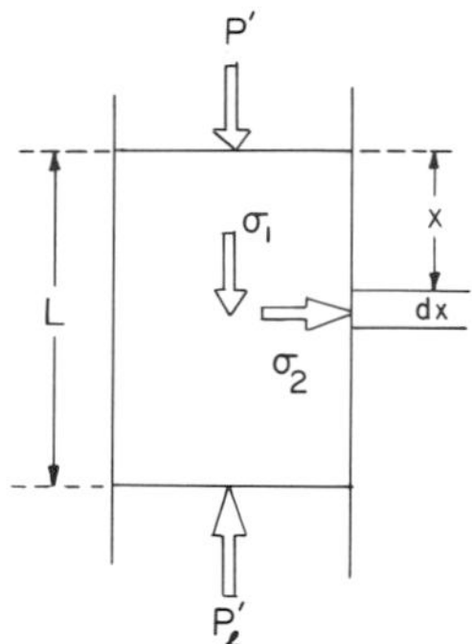

Fig. V-25 Pressures and stresses in a compact being compressed.

This is known as the Shaxby–Evans–Unckel equation (Shaxby and Evans, 1963; Unckel, 1945).

V-4-1 Lubrication Index

Frequently the lubrication index of a powder mass is taken as the ratio P_l'/P', i.e., the ratio of the lower to upper punch pressure forces (Higuchi, 1954). Guyot *et al.* (1977) have suggested that the work upon the lower punch divided by the work input of the upper punch during the compression cycle be used in place of the force ratio as a lubrication index. (See Section V-4-2.) Bowden and Tabor (1954) have pointed out that frictional force is the result of the shearing of contact points of the two surfaces sliding against one another. Lubricants constitute a layer of material of low shear strength between powder and die wall. Shotton and Hersey (1970) have reported the values listed in Table V-8. Guyot *et al.* (1977) have pointed out that curves such as those shown in Fig. V-14a always drop below Q_2. The further this deviates from the base line the worse are the sticking and lubrication problems of the material.

TABLE V-8

Shear Strength of Stearate Lubricants

Lubricant	Shear strength ($MN\ m^{-2}$)	Lubricant	Shear strength ($MN\ m^{-2}$)
Zn stearate	1.0	Mg stearate	2.0
Stearic acid	1.3	Na stearate	3.3
Ca stearate	1.5		

V-4-2 Work in Compression

It is possible to monitor the work expended by the upper punch by recording simultaneously the applied force P' of the upper punch and its displacement x. Plots such as shown in Fig. V-26a are obtained. Marshall (1977) points out that the area ABCD (denoted W_g) is the work done by the upper punch. The curve CE is a result of elastic recovery (while the tablet is still in the die) so that the cross-hatched area W_f is the energy "stored" in the tablet. It should be noted that there can still be an elastic recovery upon ejection (further reducing the energy stored), since many tablets expand upon ejection. Marshall (1977) points out that recompression of a tablet while still in the die and use of the adjusted W figures can be utilized to evaluate a tablet formulation. The figures in Table V-9 are indicative of this. Here the first column represents unlubricated granulation compressed in an unlubricated die, the second represents unlubricated granulation compressed in a lubricated die, and the third represents lubricated granulation. In other words, the three situations represent progressively better states of lubrication. Note from the first column that the punch force ratio does not distinguish between the two latter cases, whereas the net work terms in the last line do.

Marshall (1977) and (as mentioned) Guyot *et al.* (1977) have pointed out that the ratio of areas (b) to (a) in Fig. V-26 is a better measure of lubrication than the ratio of lower to upper punch force. Guyot *et al.* (1977) have also pointed out that the shapes of the curves sensed by the lower punch are frequently indicative of the type and adequacy of lubrica-

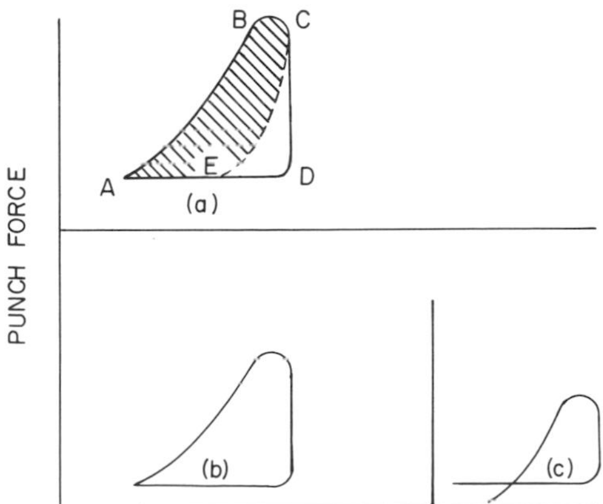

Fig. V-26 Force–displacement diagram in a compression cycle: (a) upper punch displacement; (b, c) lower punch displacement.

TABLE V-9

Force–Displacement Curves in Double Compression with Different States of Lubrication

	Unlubricated	Die wall lubricated	Granulation lubricated
P'/P'_1	0.82	0.96	0.96
Punch work, first cycle	1.20	1.10	0.90
Punch work, second cycle	0.60	0.60	0.55
Net work	0.60	0.50	0.35

tion and that, e.g., crossing of the lines, such as shown in Fig. V-26c is typical of poorly lubricated material. The displacement reference is, however, still the upper punch.

V-4-3 Temperature Rises in Compaction

The compression of tablets is accompanied by a rise in temperature. Nelson *et al.* (1955) have determined that there should be about a 5°C temperature rise during the compression of sulfathiazole tablets. Hanus and King (1968) used thermocromic indicators to study temperature rises in sodium chloride tablets and found, experimentally, increases up to 30°C. Travers and Merriman (1970) measured the time–temperature relationships *during* the compression cycle by inserting thermistors in tablets (of sodium chloride and boric acid). They utilized compression forces up to 50 kN. They found a rise in temperature during the initial compression, and, in the relaxation stage, they found that the temperature dropped. The temperature rises in the compression stage were found to be proportional to the applied pressure *P*.

Since the energy stored is what is dissipated as heat loss, the mechanical equivalent of this latter should correlate with the stored energy. Marshall (1977) has found good correlations of these two parameters.

REFERENCES

Berenbaum, R., and Brodie, I. (1959). *Br. J. Appl. Phys.* **10**, 281.

Bowden, F. P., and Tabor, D. (1967). "Friction and Lubrication," p. 110. Wiley, New York.

Brook, D. P., and Marshall, K. (1968). *J. Pharm. Sci.* **57**, 481.

Brown, R. L., and Richards, J. C. (1960). *Trans. Inst. Chem. Eng.* **38**, 243.

Carstensen, J. T. (1973). "Theory of Pharmaceutical Systems," Vol. II, p. 209. Academic Press, New York.

Carstensen, J. T., and Chan, P. L. (1977). *J. Pharm. Sci.* **66**, 1235.

Carstensen, J. T., and Laughlin, S. (1979). *Powder Technol.* **23**, 79.
Carstensen, J. T., Johnson, J. B., Valentine, W., and Vance, J. J. (1964). *J. Pharm. Sci.* **53**, 1050.
Cooper, A. R. Jr., and Eaton, L. E. (1962). *J. Am. Ceram. Soc.* **45**, 97.
Danish, F. Q., and Parrott, E. L. (1971). *J. Pharm. Sci.* **60**, 550.
David, S. T., and Augsberger, L. L. (1977). *J. Pharm. Sci.* **66**, 155.
Fell, J. T. and Newton, J. M. (1971). *J. Pharm. Sci.* **60**, 1428, 1868.
Guyot, J. C., Delacombe, A., Merle, C., Becourt, P., Ringard, J., and Traisnel, M. (1977). *Proc. Int. Conf. Pharm. Technol. 1st, Paris May 31–June 2, 1977* **IV**, 142.
Hanus, J. H., and King, L. D. (1968). *J. Pharm. Sci.* **57**, 677.
Heckel, R. W. (1961). *Trans. ASME* **21**, 671, 1001.
Hersey, J. A., and Rees, J. E. (1971). *Nature* **230**, 96.
Hiestand, E. N. (1978). *Proc. Int. Conf. Powder Technol. Pharmacy, 6–8 June, 1978, Basel, Switzerland*, p. 10.
Hiestand, E. N., and Peot, G. B. (1974). *J. Pharm. Sci.* **63**, 605.
Hiestand, E. N., Bane, J. M. Jr., and Strezelinski, E. P. (1971). *J. Pharm. Sci.* **60**, 758.
Hiestand, E., Wells, J. E., Peot, C. B., and Ochs, J. E. (1977). *J. Pharm. Sci.* **66**, 510.
Higuchi, T. (1954). *J. Am. Pharm. Assoc. Sci. Ed.* **43**, 344.
Higuchi, T., Rao, A. N., Busse, L. W., and Swintosky, J. V. (1953). *J. Am. Pharm. Assoc. Sci. Ed.* **42**, 194.
Houwink, R. (1954). "Elasticity, Plasticity and Structure of Matter," p. 2. Dover, New York.
Huffine, C. L. (1953). Ph.D. thesis. Columbia Univ., New York.
Jastrzebski, Z. D. (1976). "The Nature and Property of Engineering Materials," 2nd ed., p. 249. Wiley, New York.
Jayasinghe, S. S. (1970). Ph.D. thesis. London Univ., London, England.
Jayasinghe, S. S., Pilpel, N., and Harwood, C. F. (1969). *Mater. Sci. Eng.* **5**, 287.
Koerner, R. M., and McCabe, W. M. (1972). *Proc. 1972 Powder Metall. Conf.*, p. 225–241.
Leigh, S., Carless, J. E., and Burt, B. W. (1967). *J. Pharm. Sci.* **56**, 888.
Long, W. M. (1960). *Powder Metall.* **6**, 73.
Marshall, K. (1977). The Physics of Tablet Compression, paper presented at *Ann. Arden House Conf., 12th, Harriman, New York, February 2, 1977.*
Nelson, E., Busse, L. W., and Higuchi, T. (1955). *J. Am. Pharm. Assoc. Sci. Ed.* **S44**, 223.
Newmann, B. S. (1967). *Adv. Pharm. Sci.* **2**, 181.
Paris, J., Duchene, D., and Puisieux, F. (1975). Paper presented at the *Int. Conf. Compression, 2nd, Brighton, England, September 2–4, 1975.*
Perelman, V. E., and Roman, O. V. (1971). *J. Powder Metall.* **9**, 692.
Pollack, H. W. (1964). "Applied Physics," pp. 133–136. Prentice Hall, Englewood Cliffs, New Jersey.
Rankell, A. S., and Higuchi, T. (1968). *J. Pharm. Sci.* **57**, 574.
Ridgway, K., and Rupp, R. (1969). *J. Pharm. Pharmacol.* **21**, 30S.
Schwartz, E. G., and Holland, A. R. (1969). *Int. J. Powder Metall.* **5**, 79.
Schwartz, E. G., and Weinstein, A. S. (1965). *J. Am. Ceram. Soc.* **48**, 346.
Shaxby, J. H., and Evans, J. C. (1963). *Trans. Faraday Soc.* **19**, 60.
Shlanta, S., and Milosovich, G. (1964). *J. Pharm. Sci.* **53**, 562.
Shotton, E., and Hersey, J. A. (1970). Compaction and Compression. *In* "The theory and Practice of Industrial Pharmacy," (L. Lachman, H. A. Lieberman, and J. L. Kanig, eds.), Chap. 9, p. 183. Lea and Febiger, Philadelphia.
Skotnicky, J. (1953). *Czech. J. Phys.* **3**, 225.
Strijbos, S., Rankin, P. J., Klein, R. J., Wassink, M., Bannick, J., and Oudemans, G. J. (1977). *Powder Technol.* **18**, 187.

Takieddin, M., Puisieux, F., Didry, J. R., Touré, P., and Duchenê, D. (1977). *Int. Conf. Powder Technology, 1st, Paris, May 31, 1977*. Abstracts, Vol. I, p. 248.

Train, D. (1956). *J. Pharm. Pharmacol.* **8**, 45T.

Train, D. (1957). *Trans. Inst. Chem. Eng.* **35**, 258.

Travers, D. N., and Merriman, M. P. H. (1970). *J. Pharm. Pharmacol.* **22**, 11S.

Unckel, H. (1945). *Arch. Eisenhuettenwesen* **18**, 161.

York, P., and Pilpel, N. (1972a). *J. Pharm. Pharmacol.* **24**, 47P.

York, P., and Pilpel, N. (1972b). *Mater. Sci. Eng.* **9**, 281.

York, P., and Pilpel, N. (1973). *J. Pharm. Pharmacol.* **25**, 1P.

CHAPTER

VI

Physical Properties of Solid Dosage Forms

The final product in solid dosage form is either a powder (e.g., ViMagna granules), a capsule (e.g., Librium capsules), or a tablet.

For a *powder* the requirements are flowability, dissolution, noncaking on storage, and stability. Only *noncaking* need be discussed here, the rest having already been treated in Chapter III. In addition, there is the vague property of *physical appearance*, which also will be dealt with in general.

For *capsules* important physical properties are *disintegration* and dissolution, first *dissolution of the gelatin shell*, then of the contents. The former creates a lag time; the latter either is a powder (cube root) dissolution or (if caking has taken place) a dissolution akin to that of tablets. In the case of capsules (both hard and soft shell) the *fragility* of the capsule is of importance, since a broken capsule in a bottle produces a nonsalable item.

For tablets, the friability, disintegration, and dissolution are important characteristics. Although solid state stability has been discussed for particulate solids, the *stability of tablets* (and caked capsules) differs from that of powders and will be discussed separately. The *dissolution of tablets* and capsules is also different from that of powders.

Hence this chapter is subdivided into seven parts:

(1) powder caking,
(2) the dissolution of gelatin shells,

(3) the fragility of gelatin shells,
(4) tablet and plug disintegration,
(5) tablet dissolution,
(6) the stability of solid dosage forms, and
(7) physical appearance.

VI-1 POWDER CAKING

The caking of powder in bulk (e.g., in drugs) is frequently due to cohesion C and electrostaticity. If a freshly milled powder is poured into a drum, it will at first be free flowing. It will rest in the drum at its cascaded apparent density. On storage it will, however, "consolidate."† There are several ways of following the development of this.

In a fresh powder the volume reduction rate constant such as described by Carstensen (1973) and Takieddin *et al.* (1977) can be followed. This, however, differs from the slow changes occurring on storage in a bulk powder. Here, granted, volume reduction also takes place, but the important factor is the *caking*. This term, the meaning of which is qualitatively self-evident, is a result of *bonding* and occurs by mechanisms (approach of surfaces and proper alignment) of the same type as those described for direct compression (Fig. IV-25). Plastic deformation may also occur in the parts of a drum which are under pressure (at or near the bottom) but is not quite as likely to occur either in a powder in a bottle or in a capsule. It should be pointed out that moisture (liquid bridging) plays an important part in caking.

Note that caking is a time-dependent phenomenon. To test it in semiquantitative fashion the following methods have been used:

(1) storing the powder in cylinders with weights on top and measuring the volume as a function of time, and

(2) performing this type of experiment at various humidities.

The volume itself is at best a pseudoquantitative measure of the caking σ. Better is a tensile-strength or other hardness test of the compact utilizing an equation such as Eq. (V-3-15). In this manner σ can be measured as a function of time. When bonding is not strong but is still significant, a friability or a milling test can be employed. Finally, a sieve analysis can be used and the increase in particle size as a function of time taken as a measure of increased cohesion.

In some cases the caking is *chemical*. In the simplest case the eutectic

† This use of the word differs from its usage during tableting.

temperature is above the storage temperature. This, in the most unfortunate proportion, will cause liquefaction (Carstensen, 1964, 1973, 1977) but in general, if 5% liquid or less is formed, then only a liquid layer will form. This will aid in consolidation and bond formation. The relative amounts of liquid and solid of course can be found by the weight arm rule as described in Chapter I.

Another type of chemical interaction occurs with the capsule shell. Certain hydrochloride salts (particularly dihydrochloride) are very hygroscopic and will extract water from the shell of a gelatin capsule, even to the extent that the moisture and hydrochloric acid transform such substances as magnesium stearate into magnesium chloride and stearic acid. This in turn promotes cake formation.

VI-2 DISSOLUTION OF GELATIN SHELLS

For hard-shell capsules it should be noted that dissolution in water at 25°C is fairly slow. Hydrochloric acid and (in particular) enzymes (at 37°C) make dissolution rapid. In certain cases (aldehydes) there is an interaction with the gelatin (amine–aldehyde condensation, giving —N= CH_2—), which slows the dissolution rate down and (particularly with formaldehyde) makes the gelatin insoluble.

In the manufacture of soft-shell capsules the film is considerably thicker than in hard shells and dissolution is a function of film thickness. Formulation affects the dissolution process as well. Iron oxides (which are used as pigments), particularly on reuse (remelting) of unused gelatin "web," give rise to hardening. Moisture content is also a factor, although no published data exist for this parameter. Dissolution, dilution, and removal of dissolved gelatin by the movement of the dissolution medium are obviously factors which affect the dissolution rate of a soft gelatin capsule form. It should be pointed out that in the USP basked dissolution method, the dosage form is placed in a basket consisting of a 40 mesh screen, and in this test the screen can be plugged up by gelatin, in the dissolution both of hard and of soft shells. In one-compartment methods (such as the USP paddle method) the capsules may float and are therefore usually placed in a copper coil.

VI-3 FRAGILITY OF GELATIN SHELLS

Hard-shell capsules are two-piece capsules. The powder in the capsule acts as a dash pot, buffers impacts, and renders the capsules nonfragile.

Bonding or spot welding somewhat reduces this elasticity, but not to the point where it is a problem.

Soft-shell capsules, on the other hand, are one piece and impacts are totally absorbed by the shell. If this contains insufficient moisture, then the capsule will crack, giving not only a defective capsule, but also a defective bottle of capsules, since either the liquid or the solid on the inside of the capsule can then find its way out of the capsule to the surface of neighboring units.

Common are *fall tests*, where the capsules in a container are let drop a certain distance. A narrow range A_1–A_2 of moisture content (in milligrams per capsule) is aimed at, since below A_1 the capsules will be fragile and above A_2 they will deform on storage. The range from A_2 to A_1 is frequently less than 10 mg per capsule.

VI-4 TABLET AND PLUG DISINTEGRATION

Disintegration is important when the dosage form in contact with biological fluids does not immediately fall apart into distinct particles. This at times happens to capsule fills, which may form plugs when they contain too much magnesium stearate or when solid interactions occur on storage. It is the exception for capsules, but tablets, in general, take finite times to disintegrate and must disintegrate rapidly in order to provide rapid blood levels. To this end, disintegrants are added. The most notable disintegrants are cornstarch, alginic acid, and modified cornstarch. There are two types of the latter: (a) gelatinized starches (e.g., StaRx®) or (b) chemically modified starches, e.g., sodium starch glycolate (Primojel®). Disintegration depends on two factors: (a) penetration of water (disintegration medium) into the tablet or plug and (b) swelling of the disintegrant after contact with water (Couvreur, 1975). Considering the tablet to be porous with *one* (average) pore radius r (which is obviously an approximation) allows expression of the rate of moisture penetration as (Washburn, 1921; Nogami *et al.*, 1966; Couvreur, 1975)

$$dL/dt = Qr^2/8\eta L \tag{VI-4-1}$$

where η is the viscosity of the liquid, L is the length of penetration, and

$$Q = (2\gamma \cos\theta/r) - g\rho \sin\alpha \tag{VI-4-2}$$

Here γ is interfacial tension, θ is the contact angle, g is the gravitational acceleration, ρ is the density of the liquid, and α is the angle between liquid and capillary wall. From Eq. (VI-4-1) it follows (Couvreur, 1975) that

$$L^2 = (r\gamma \cos\theta/2\eta)t \tag{VI-4-3}$$

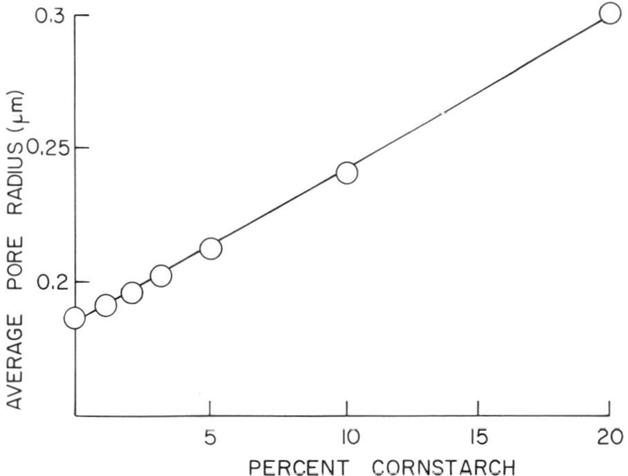

Fig. VI-1 Average pore radius of an Emcompress® tablet as a function of percent of added cornstarch. [After Couvreur (1975).]

The amount of starch (or modified starch) affects the medium pore radius (Fig. VI-1) and the contact angle (Fig. VI-2).

Disintegrants are classified [after Couvreur (1975)] into seven types:

(1) starches,
(2) natural gums (e.g., arabic, guar),
(3) cellulosic derivatives (methyl cellulose, sodium carboxymethyl cellulose),
(4) microcrystalline cellulose,
(5) alginic acid,

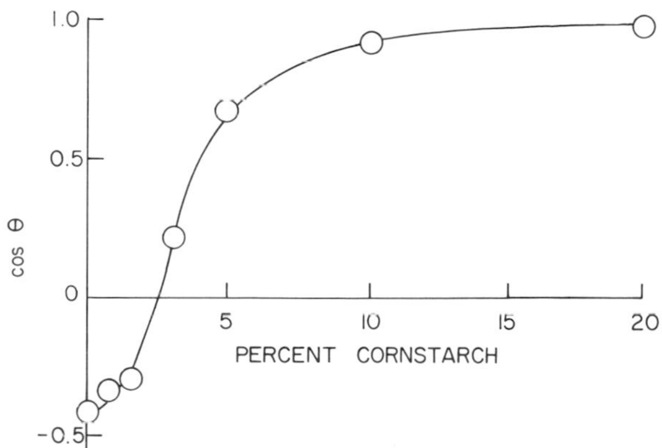

Fig. VI-2 The cosine of the contact angle as a function of the percentage of cornstarch in an Emcompress® tablet. [After Couvreur (1975).]

(6) aluminum silicates (bentonite, montmorillonite), and
(7) enzymes.

The mode of disintegration has been thought of as due to one of three causes:

(a) swelling of the disintegrant (the expansion in turn breaking the tablet mechanically),
(b) penetration of liquid into the tablet [i.e., a phenomenon connected with porosity and increased penetration rates such as shown in Eq. (VI-4-3)], and
(c) the heat of interaction of the disintegrant with the dissolution fluid (water) (the increased temperature causing bonds to break).

Disintegration was originally thought to be due solely to the first of these (i.e., to swelling). Ingram and Lowenthal (1966, 1968) measured the maximum swelling of starch in water and found it to be 33% by volume at 37°C. They reasoned that the swelling should account for at least the equivalent of the pore volume (and reasoned that this might be why a minimum amount of starch is always necessary), but since porosity is mostly in excess of 5%, at least 16% starch would be necessary for disintegration, and it appears that other mechanisms are important as well.

In general both penetration and swelling must occur. The beneficial effects of starch (and other disintegrants) in penetration are described in Eq. (VI-4-3). If the situation is as shown in Fig. VI-3 then the starch particle must expand from R to $R + 2r$ in radius. The 33% volume expansion corresponds to a 10% expansion in radius, so that only pores with radius r' contribute to disintegration, where $R + 2r' = 1.1R$; i.e.,

$$r' = 0.05R \qquad \text{(VI-4-4)}$$

is the maximum useful pore radius.

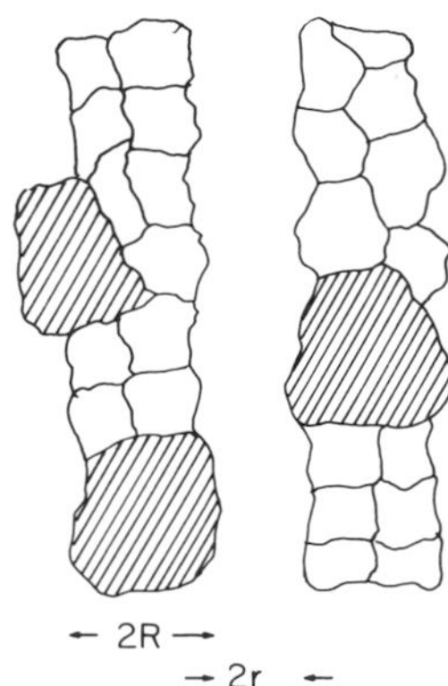

Fig. VI-3 Schematic showing the role of the diameter of starch (cross hatched) and the pore diameter in tablet disintegration.

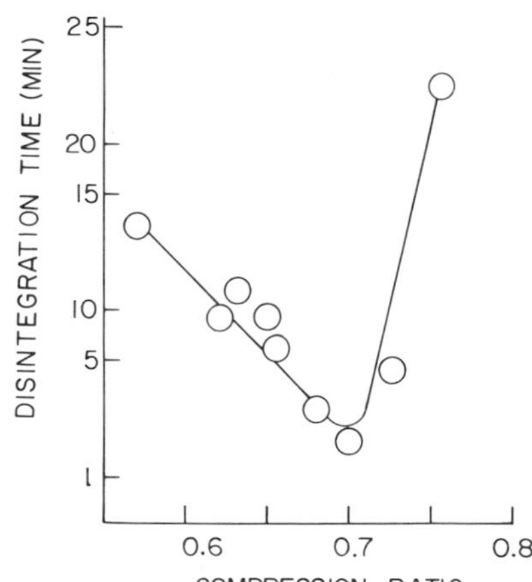

Fig. VI-4 Disintegration time as a function of compression pressure. [After Berry and Ridout (1950).]

In general pore size r decreases with applied tablet compression pressure, and most often [e.g., Kennon and Swintosky (1958)] disintegration times also increase with applied pressure. This is because r in Eq. (VI-4-3) becomes smaller, so that penetration rates decrease; however, at very low pressures r is too large, larger than r' in Eq. (VI-4-4), so that disintegration times at sufficiently low pressure will increase with decreasing applied pressure [see Berry and Ridout (1950) and Fig. VI-4]. Khan and Rhodes (1976) have described a similar decrease with pressure in some direct compression formulas.

Carstensen (1976) and Carstensen *et al.* (1978a, p. 48) have pointed out that most tablets swell when they disintegrate, and that equations describing disintegration in terms of the thickness of the tablet as a function of time are less practical than those describing the amount (mass) of material not disintegrated at time t on an anhydrous basis. They determined this experimentally (Carstensen *et al.*, 1978a, p. 982) and found the mass to decrease in time mono- or biexponentially, as shown in Fig. VI-5. One may consider a tablet as consisting of T_0 particles (i.e., in the dissolution liquid it will break up into T_0 particles). Khan and Rhodes (1975) have pointed out that these are not all the same size (but are distributed, e.g., lognormally) and that the size distribution differs from that of the granulation or powder (and that this, in fact, is a measure of the bonding rendered by the compaction).

From the point of view of constructing a model, the situation may be simplified by assuming that the T_0 particles are identical and that the number of particles *not* disintegrated at time t (i.e., that are still in the intact tablet) is given by

$$T = T_0 \exp(-qt) \tag{VI-4-5}$$

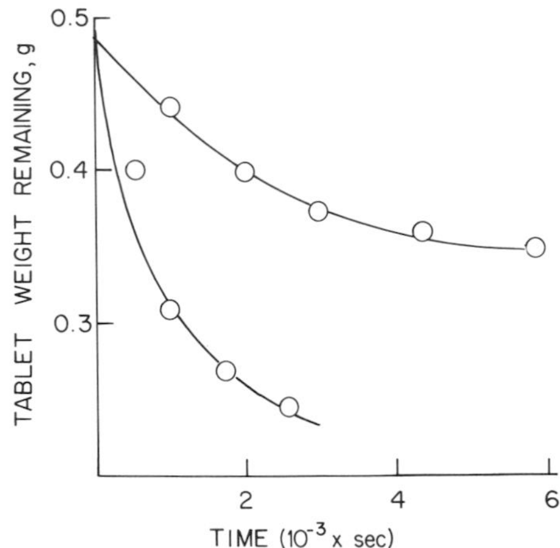

Fig. VI-5 Weight versus time curves of directly compressed tablets. Top curve, no starch; bottom curve, 50 mg of starch per tablet. [After Carstensen *et al.* (1978a, p. 982). Reproduced with permission of the copyright owner, the *Journal of Pharmaceutical Sciences*.]

where q is a disintegration rate constant. Kitamori and Shimamoto (1976) and Kitamori and Iga (1978) have used a formula for the number of particles N released:

$$N = T_0/(t/t_d)^m \tag{VI-4-6}$$

where t_d is the disintegration time and m a constant. Equations (VI-4-5) and (VI-4-6) are comparable and equally good input and descriptive functions for the disintegration process. Nelson and Wang (1977, 1978) have used dissolution rate data to back-calculate disintegration functions.

VI-5 DISSOLUTION OF DOSAGE FORMS IN DISSOLUTION APPARATUSES

In developing a drug dosage form, clinical batches are made and tested in patients. These batches may have varying sizes and eventually form the basis for the production formula and procedure used. Since it is not possible to test clinically each production batch, *in-vitro* parameters of importance in the clinical batches are sought and correlated with the *in-vivo* performance of the batches. With such a correlation, future batches can be tested for the *in-vitro* parameter of correlation and if it is within the range found in the batches used in the clinic it may be concluded that the production batch is "bioequivalent" with the clinical batches upon which product claims were founded. The most common parameter of this sort is the dissolution rate of the tablets (or capsules).

In a dissolution apparatus, dissolution is effected by (1) moving the liquid in relation to the (stationary) dosage form or (2) moving both. Apparatuses where the liquid is static and the dosage form moves, although plausible, have not been developed. Type (1) is called a flow apparatus and type (2) will be referred to simply as a nonflow method.

Flow apparatuses are common in Europe (Langenbucher, 1969; Kohler *et al.*, 1975; Bathe *et al.*, 1975; Tingstad and Riegelman, 1970). Here the dosage form is placed in a chamber and liquid let flow through.

If Eq. (VI-4-5) or (VI-4-6) holds, then the dissolution rate is given by that of the nondisintegrating tablet plus that of the dislodged particles (Carstensen *et al.*, 1978a, p. 48). At time t the number of dislodged particles is given by

$$T = T_0[1 - \exp(-qt)] \tag{VI-5-1}$$

as long as none have dissolved completely. Assuming a cube root law, where the diameter d is a function of time, solubility S, intrinsic dissolution rate constant k, density ρ, and initial diameter d_0 (Carstensen, 1977), then

$$d = d_0 - (2Sk/\rho)t = d_0 - Kt \tag{VI-5-2}$$

where $K = 2Sk/\rho$ and where the time period referred to is given by

$$0 < t < d_0/K \tag{VI-5-3}$$

At time t, the amount of particles having been born in time τ is

$$(\partial T/\partial t)_\tau \, d\tau = T_0 q \exp(-q\tau)\, d\tau \tag{VI-5-4}$$

so that they constitute a mass of undissolved material of

$$T_0 q \exp(-q\tau)\, d\tau \,[d_0 - K'(t-\tau)]^3(\pi/6) \tag{VI-5-5}$$

where K' is the cube root dissolution rate constant ($K' = Km_0^{1/3}/d_0$). Hence the total mass undissolved is

$$m = T_0 \exp(-qt)(\pi/6)d_0^3 + \int_0^t T_0 q \exp(-q\tau)[d_0 - K'(t-\tau)]^3(\pi/6)\, d\tau \tag{VI-5-6}$$

In the time period $t > d_0/K$ there are particles born before times $\tau > d_0/K$ which are completely dissolved and those born before times $\tau < d_0/K$ which contribute in the same sense as described by Eq. (VI-5-6). Hence, in this case, the lower limit in the integral is changed from zero to d_0/K.

The integral in Eq. (VI-5-6) can be solved in closed form, since

$$\int (a - bx)^3 \exp(-qx)\, dx = -b^3 \int x^3 \exp(-qx)\, dx + \text{lower terms} \tag{VI-5-7}$$

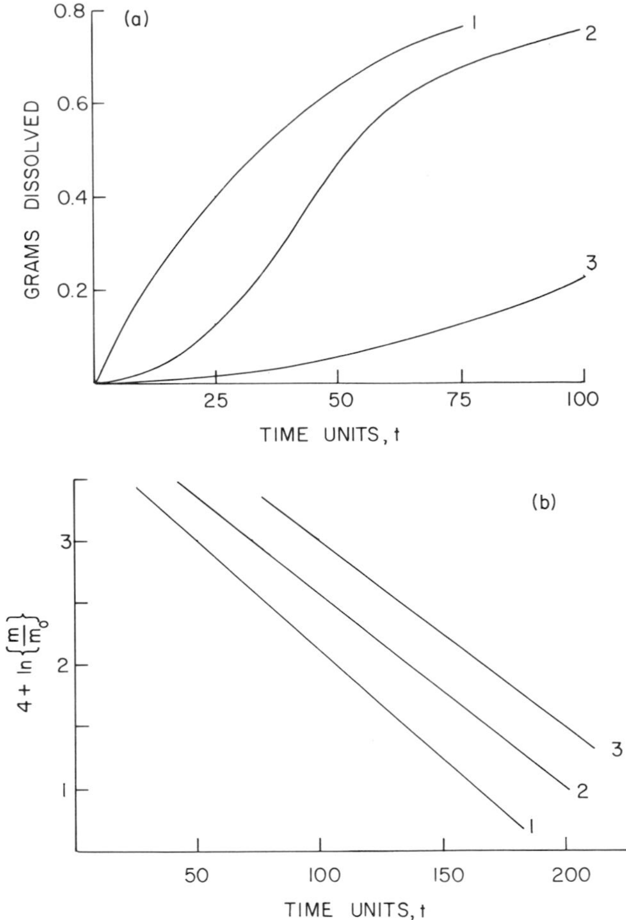

Fig. VI-6 (a) Dissolution in a basket apparatus where the linear liquid velocity is twice as large in the basket as outside. Tablet weight, 1 g. The disintegration constant is $q = 0.02$ reciprocal time units. Values of $2kS/\rho$ are (curve 1) 0.15, (curve 2) 0.0015, and (curve 3) 0.00015 cm per time unit. (b) The data in (a) plotted according to Eq. (VI-5-8). [After Carstensen *et al.* (1978a, p. 48). Reproduced with permission of the copyright owner, the *Journal of Pharmaceutical Sciences*.]

and the integral $-b^3 \int x^3 \exp(-qx)\, dx$ can be solved using integration by parts. It is then possible to generate dissolution curves for various values of q and K. These give rise to S-shaped curves like the ones shown in Fig. VI-6 (Carstensen *et al.*, 1978a, p. 982).

The third type of dissolution apparatus is a two-compartment apparatus (e.g., the USP basket apparatus). Here there are two dissolution compartments with different hydrodynamics. As will be discussed shortly, the dissolution rate constant k (cm/sec) must be a function of the liquid velocity of the dissolution medium. Carstensen *et al.* (1978a, p. 48) have approximated this by assuming two average velocities, one in the basket and one outside [and in effect established these velocities experimentally (Carstensen *et al.*, 1978b)]. This then gives three time periods:

$0 - \theta_1$: tablet + particle in basket;
$\theta_1 - \theta_2$: (particle + tablet in basket) + particle outside basket;
$t > \theta_2$: (particle + tablet in basket) + particle outside basket + dissolved particles.

Again these give rise to S-shaped curves, which can be approximated by

$$\ln m = \ln m_0 - k'(t - t_i) \qquad \text{(VI-5-8)}$$

where k' is a "dissolution constant" and t_i is a lag time. Carstensen *et al.* (1978a, p. 48) have shown that the lag times t_i for a large range of q and K values are log–log related to the K values, and have found experimentally (Carstensen *et al.*, 1978a, p. 982) that, in a large range of values of K and q, q and k' are log–log related (Figs. VI-7 and VI-8). These studies, as well as those of Nelson and Wang (1977, 1978) and of Kitamori and Iga (1978) then give a rational explanation for the S-shaped curves that are frequently observed in the dissolution of dosage forms in dissolution apparatuses. Attempts to approximate these by probit functions (Wagner, 1969) and by Weibull functions (Langenbucher, 1972) may at times be successful but at

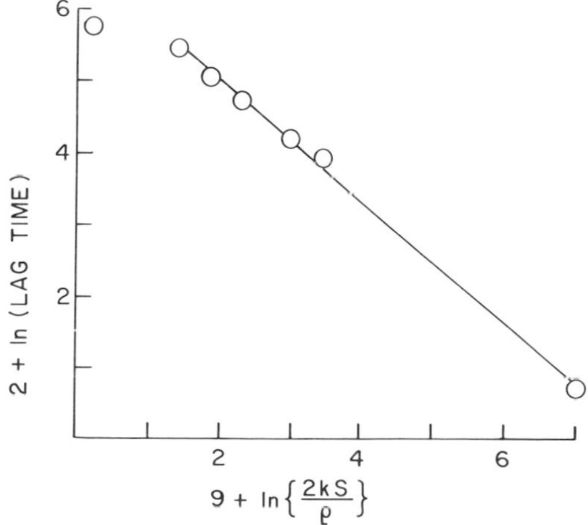

Fig. VI-7 Lag times t_i obtained from curves such as shown in Fig. VI-6b, for seven values of $2kS/\rho$. [After Carstensen *et al.* (1978a, p. 48). Reproduced with permission of the copyright owner, the *Journal of Pharmaceutical Sciences*.]

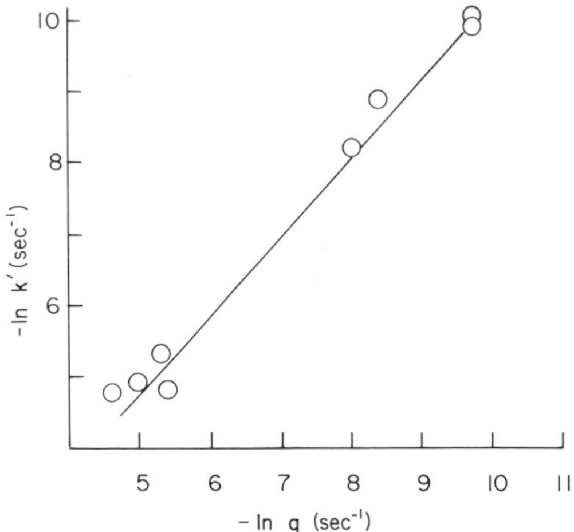

Fig. VI-8 Experimental data from tablets (either directly compressed or wet granulated) containing different amounts of cornstarch, showing correlation between ln k' and ln q. [After Carstensen *et al.* (1978a, p. 982). Reproduced with permission of the copyright owner, the *Journal of Pharmaceutical Sciences*.]

other times do not result in a good fit, and they are, at best, phenomenological. A more rational approach uses k' and t_i in equations of the same type as Eq. (VI-5-8). It is interesting to note that in the large ranges of K and q values referred to, k' is a function of the disintegration constant (as is t_i) and that then the *dissolution is a function of disintegration in this range*. A disintegration (rather than a dissolution) rate test is therefore much more appropriate for products that are of this type. This is not to say that the dissolution test serves no purpose, since a bad batch could (e.g., through a decrease in surface area) cause a drop in the value of K to such an extent that the time course of dissolution would, indeed, again become dictated by the dissolution rate.

There have been many problems in dissolution testing from the point of view of *reproducibility*. It is particularly the laboratory-to-laboratory consistency that is lacking, and this is presumably owing to two main factors: (a) There are different vibrational levels at different locations, giving rise to different disintegration rates of the tablets, and (b) the hydrodynamics of the apparatuses may vary because of the different locations of the various components of the set-up (small differences in alignment, for instance).

To study the latter point, Carstensen and Dhupar (1976) and Carstensen *et al.* (1978b) studied the dissolution of oxalic acid as a function of the linear (laminar) velocity v in a flow apparatus. A note on the anticipated effect of v on k (both in centimeters per second) is in order at this point.

One theory of dissolution is the so-called film theory [reviewed, e.g., by Carstensen (1974, 1977)]. This theory assumes that the particles are surrounded by a film of smaller (and circular) velocities in which the concentration rises from the bulk concentration C at the boundary between film and bulk liquid up to S at the crystal surface (where the velocity is zero). If the film thickness is h, then Fick's first law yields

$$\text{Flux} = (1/A)(dm/dt) = -D(dC/dx) = -D[(C-S)/h] \quad \text{(VI-5-9)}$$

(if the gradient dC/dx is linear). Here D is the diffusion coefficient of the dissolved species and A the surface area of the solid. Hence

$$dm/dt = (D/h)A(S-C) \quad \text{(VI-5-10)}$$

and this is identical with Eq. (II-5-1) provided that the intrinsic dissolution rate constant k (cm/sec) is equal to

$$k = D/h \quad \text{(VI-5-11)}$$

It may therefore be seen that the intrinsic dissolution rate "constant" is a function of the film thickness h. It is known from hydrodynamic theory (Levich, 1962) that the film thickness is a function of the liquid velocity by a relation of the type

$$h = fv^{-n} \quad \text{(VI-5-12)}$$

where the parameter n is about $\frac{1}{2}$. Introducing this into Eq. (VI-5-11) then

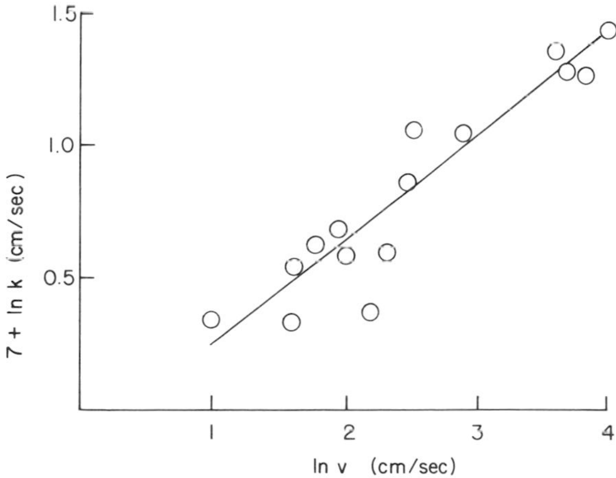

Fig. VI-9 Dissolution rate constants k (cm/sec) as a function of liquid velocity (cm/sec). Part of the points are from a column apparatus, part from a USP basket dissolution apparatus. The test substance is oxalic acid. The least-squares fit is $\ln k = 0.63 \ln v - 7.53$. [After Carstensen *et al.* (1978b). Reproduced with permission of the copyright owner, the *Journal of Pharmaceutical Sciences*.]

gives

$$k = fDv^n \tag{VI-5-13}$$

Carstensen and Dhupar (1976) and Carstensen *et al.* (1978b) indeed found this to be true (Fig. VI-9) and utilized the knowledge that for oxalic acid

$$k = 5.4 \times 10^{-4} v^{0.63} \tag{VI-5-14}$$

to determine the hydrodynamics of several dissolution apparatuses.

VI-6 IN VITRO–IN VIVO CORRELATIONS

As mentioned earlier, the main importance of dissolution tests is their being *control tools*. When a drug is developed, as described in the introduction to Section VI-5, batches are made for clinical testing. It is natural to attempt to make these batches as good as possible, e.g., to have as good a dissolution as possible. If it is assumed that Eq. (VI-5-8) holds and it is established that the fraction F of drug absorbed by the patient is the biological parameter which correlates with in vitro dissolution, then, e.g., six clinical batches would give in vitro half lives of $t_1, t_2, \ldots, t_6$. These include the lag time and are of the form

$$t_{1/2} = (0.693/k') + t_i \tag{VI-6-1}$$

If the corresponding areas under the blood level curves (or peak blood levels or some other pertinent parameter) are denoted $B_1, B_2, \ldots, B_6$, then the correlation will take the form shown in Fig. VI-10a. Plotting techniques, in general, dictate offsetting the scales and scaling them so that a line sloping about 45° appears with a "spread-out" abscissa on the graph. In such cases, however, an important feature is lost, and if the origin is included in the graph (i.e., making the scaling somewhat poor), as shown in Fig. VI-10b, it may be seen that correlation can never be good, because the spread of the abscissa points is too narrow. From a statistical point of view it would be advantageous to test a poor batch or two (points P and Q in Fig. VI-10b) so that a meaningful regression line could be obtained. This would aid in setting good specifications (most specifications being too narrow and frequently giving unnecessary production problems).

In improving dissolution rates, fine milling is often resorted to. Here excessive milling may cause agglomeration (as may electrostatic charges) and this may be counterproductive. Another point which should be mentioned here is the following (Lippold, 1977): If the controlling factor in the bioavailability of a drug is a "biological window," then the amount dissolved during the time t_1 that the dosage form spends in this window is what is important. If the dissolution is by a cube root law, then (Carsten-

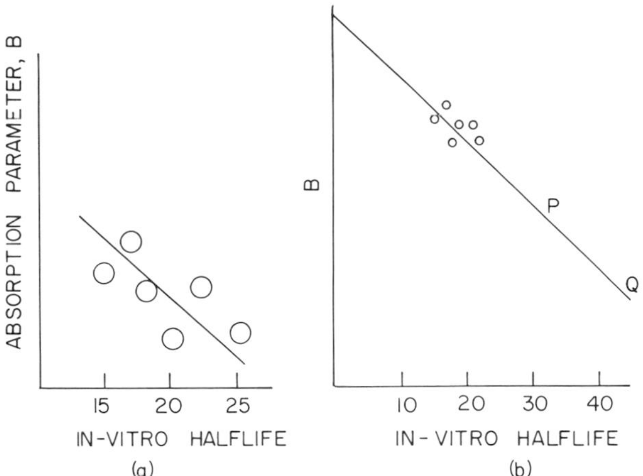

Fig. VI-10 Hypothetical in vitro–in vivo correlation data (a) as usually presented and (b) showing the clustering of experimental points that makes correlations uncertain.

sen, 1977)

$$m_0^{1/3} - m^{1/3} = \left(2kSm_0^{1/3}/\rho d_0\right)t_1 \qquad \text{(VI-6-2)}$$

which may be rewritten in terms of F as follows: The amount dissolved in the biological window is $m_0 - m$, so that the fraction absorbed is $(m_0 - m)/m_0$. Introducing this into Eq. (VI-6-2) gives

$$1 - (1 - F)^{1/3} = (2kt_1S/\rho)(1/d_0) \qquad \text{(VI-6-3)}$$

Carstensen (1977a, p. 106) has treated data by Prescott *et al.* (1970) in this fashion and shown that $(1 - F)^{1/3}$ is linear in $1/d_0$ with intercept unity. Lippold (1977) has demonstrated that it serves no purpose to reduce the particle size below the critical diameter d_c where $F = 1$, i.e., below

$$d_c = 2kt_0S/\rho \qquad \text{(VI-6-4)}$$

VI-7 SUSTAINED RELEASE DOSAGE FORMS

Sustained release dosage forms are one type of dosage form where dissolution rates become almost process dictating. The most important types are

(1) coated beadlets,
(2) erosion tablets (or beadlets),
(3) insoluble matrices,

(4) osmotic pumps, and
(5) gel formers.

In *coated beadlets* the drug is placed inside a small sphere of an insoluble organic polymer. This can be done by coating sugar beads, so-called nonpareil seeds, with a solution or suspension of the drug in sugar syrup, and then coating the particle with the polymer. It can also be accomplished by simply coating a granule of the drug with the polymer.

In any event, the process of dissolution consists of (Carstensen, 1973) (1) penetration of liquid through the film with a rate constant which depends on the thickness l of the film, then (2) dissolution of the active ingredient, and finally (3) diffusion of the dissolved drug back into the bulk solution (with a rate constant which is again a function of the polymer and of its thickness). If processes (1) and (2) take time τ, then Fick's law yields

$$\ln(C/C_0) = -AD(t-\tau) \tag{VI-7-1}$$

Here D is a diffusion rate coefficient through the film (and inversely proportional to l), C is the concentration in the solution inside the beadlet, which has a volume V equal to the volume of the uncoated beadlet (nonpareil plus drug), and C_0 is the initial concentration (i.e., VC_0 is the drug content per bead). The amount released per bead at time t is

$$V(C_0 - C) = VC_0\{1 - \exp[-AD(t-\tau)]\} \tag{VI-7-2}$$

For n_1 seeds the amount released is

$$Q_1 = n_1 VC_0\{1 - \exp[-AD(t-\tau)]\} \tag{VI-7-3}$$

If a second part is given two coats (double the film thickness) of polymer, then

$$Q_2 = n_2 VC_0\{1 - \exp[(-AD/2)(t-2\tau)]\} \tag{VI-7-4}$$

since the apparent diffusion rate is inversely proportional to the thickness and the lag time directly proportional to it. In general the release from a mixture of beadlets of p different coat thicknesses (all multiples of the original thickness) is

$$x = VC_0 \sum_i n_i\{1 - \exp[-(AD/i)(t-i\tau)]\} \tag{VI-7-5}$$

Carstensen (1973) has shown that these curves are S-shaped, and that, of course, desired characteristics can be obtained by suitable mixing of various fractions with varying film thicknesses.

For coated granules, the granulation is assumed to be log-normally distributed in size. It has been shown in Chapter IV that there is some difference in granulation between amounts of the starch paste "received" by the different granule sizes, but in the case of coating (which is sprayed on in a coating pan for instance) all sizes receive fairly much the identical amount of polymer, so that the smaller granules have a thicker coat and the larger granules have a thinner coat.

In both processes the governing step is a set of diffusion processes so that, akin to Eq. (VI-5-8), the release equation becomes

$$\ln m = \ln m_0 - k'(l)[t - t_i(l)] \qquad \text{(VI-7-6)}$$

Erosion tablets or *beadlets* are usually wax matrices containing the drug in suspended form. The wax is either melted, cooled, and milled or (for the case of tablets) a mixture of drug and milled wax is compressed. In the dissolution process the wax peels off (almost as layers or flakes) and does so in a uniform fashion, so that the release is akin to the dissolution from a crystal; i.e., in this case a cube root law applies.

The general mode of producing *insoluble matrix* preparations is (a) to granulate the drug with a solution of the (required or selected) water insoluble polymer and to remove the solvent by drying or (b) to mix the polymer and the drug and to direct-compress the mixture. In dissolution, the water will penetrate the matrix (a distance h in time t) and dissolve the drug, and the dissolved drug will diffuse out into the bulk solution. The principle governing the release from insoluble matrices is that described by Higuchi (1963), the so-called Higuchi square root law

$$Q = [2DS\epsilon(A - S\epsilon/2)]^{1/2} t^{1/2} \qquad \text{(VI-7-7)}$$

Here Q is the amount of drug released per square centimeter of surface, S is the solubility, A is the drug content in grams per cubic centimeter, and ϵ is the porosity. Fessi *et al.* (1979) have shown that the adherence of the dissolution data to Eq. (VI-7-7) stops when the tablet has just been penetrated completely and that the terminal part (usually the last 20–25%) should be diffusion controlled, i.e, adhere to Eq. (VI-5-8), where $t_i = t'$ is the time required for the tablet to fill up with water. It has also been pointed out (Fessi *et al.*, 1980) that the porosity in Eq. (VI-7-7) is not that of the matrix as produced, but ϵ_0 plus the volume fraction A' occupied by the active ingredient. The dissolution profile may therefore be summarized as

$$Q = \{2DS(\epsilon_0 + A')[A - 0.5S(\epsilon_0 + A')]\}^{1/2}(t - t_i)^{1/2}, \quad 0 < t < t' \qquad \text{(VI-7-8)}$$

$$\ln[m/m(t')] = -k'(t - t'), \qquad t > t' \qquad \text{(VI-7-9)}$$

Here t_i is the time it takes for penetration to commence, t' as mentioned is the time required for the tablet to become completely penetrated, and $m(t')$ is the amount of drug not dissolved at this time.

In the osmotic pump (Theeuwus, 1975) the drug is on the inside of a spherical, semipermeable membrane (so that the appearance is like that of a beadlet). The film has a small hole (a delivery orifice), which will allow the dissolution medium to diffuse into the core osmotically. The excess pressure will then force the solution out of the hole (assuming that the interior volume V is constant). If the surface area of the membrane is A, its thickness h, and its permeability coefficient L, then the volume flux dV/dt over the film is a reverse osmosis and is given by (Theeuwus, 1975)

$$dV/dt = (A/h)L(\gamma\Pi - P) \quad \text{(VI-7-10)}$$

Here Π and P are osmotic and hydrostatic pressures, respectively, and γ is a reflection coefficient. After an initial lag time, the delivery rate becomes constant:

$$m/m(t_i) = 1 - S/\rho \quad \text{(VI-7-11)}$$

where S is solubility.

Sustained release products can, as mentioned, also be based on *gel-forming substances* (gums). In this type of preparation the tablet (most often made by direct compression) will, when placed in a dissolution medium, acquire a layer of gel on the outside. The process of drug release thereafter consists of (a) penetration of water through the gel, (b) dissolution of the active ingredient, and (c) back diffusion of the active ingredient through the gel layer. This means that there are two diffusional processes through a barrier of time-dependent thickness. Bamba *et al.* (1979a, b) have shown that in this type of preparation Eq. (VI-5-8) holds. Here k' is a function of both the concentration of gum and the active ingredient.

VI-8 SUMMARY OF TABLET DISSOLUTION PROFILES

It is seen from the foregoing that a tablet (or capsule) may either disintegrate very rapidly (t_i negligible) or not so rapidly (t_i not negligible). In the latter case S-shaped curves occur, and if the aggregates formed on disintegration allow rapid dissolution (as compared to the disintegration), then

$$\ln(m/m_0) = -k'(t - t_i) \qquad (t_i \text{ not small; soft aggregate}) \quad \text{(VI-5-8)}$$

If the aggregates are not very permeable, then they dissolve by a square root law [Eq. (VI-7-8)] and

$$m_0 - m = A\Omega(t - t_i)^{1/2} \qquad (t_i \text{ not small; hard aggregate}) \quad \text{(VI-8-1)}$$

where Ω is the square root dissolution rate constant in Eq. (VI-7-8), and A is the surface of the aggregates formed on disintegration.

If the disintegration is rapid then

(a) if the aggregates are rapidly permeable or are the prime particles then dissolution is by the cube root law

$$m^{1/3} = m_0^{1/3} - K't \qquad \text{(VI-8-2)}$$

(b) if the aggregates are less rapidly permeable [e.g., wet granulated tablets, where the granules act as "sponges" from which the dissolved drug diffuses (Carstensen, 1974, 1977b)], then

$$\ln(m/m_0) = -k'(t - t_i) \qquad \text{(VI-5-8)}$$

where t_i is the lag time for water penetration of the granule; and

(c) if the aggregates are slowly permeable, then dissolution is by a square root law:

$$m_0 - m = -A\Omega(t - t_i)^{1/2} \qquad \text{(VI-8-1)}$$

where t_i is the time required for the water penetration to start.

Example VI-8-1 A tablet gives the following dissolution data: 5 min, 10% dissolved; 10 min, 47%; 15 min, 70%; 20 min, 83.5%; and 25 min, 91% dissolved. What type of dissolution is taking place?

Answer The data are calculated and tabulated according to Eq. (VI-5-8) in Table VI-1. The m/m_0 values are listed in column 3 and their logarithm in column 4. The least-squares fit equation of $\ln(m/m_0)$ versus time is

$$\ln(m/m_0) = -0.11834t + 0.5587$$

(The number of significant figures in both cases should only be three, but for the purpose of the following calculations five significant figures are retained.) To test the goodness of fit, the values of m corresponding to the experimental t values are calculated and are denoted $\tilde{m}$ in the table

TABLE VI-1

Treatment of Dissolution Data in Example VI-8-1 by Eq. (VI-5-8)

Time (min)	Percent released	Percent not released	$\ln(m/m_0)$	$\tilde{m}$	Δ		Δ^2
5	10	(90)					
10	47	53	−0.635	53.5	0.5		0.25
15	70	30	−1.204	29.6	0.4		0.16
20	83.5	16.5	−1.802	16.4	0.1		0.01
25	91	9	−2.408	9.07	0.07		0.00
						Total	0.42

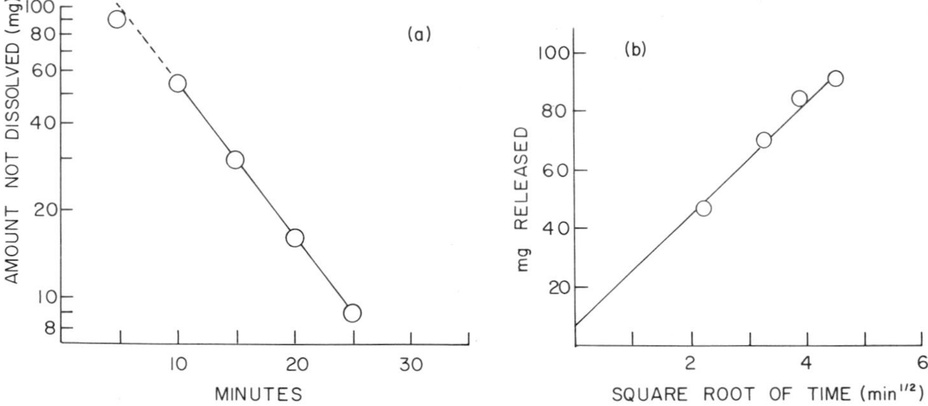

Fig. VI-11 (a) Data from Table VI-1, treated by Eq. (VI-5-8), showing m in logarithmic scaling. The lag time t_i is given by the point on the line where the abscissa has the value of m_0 (= 100). (b) Data from Table VI-1 and VI-2 treated via Eq. (VI-8-1). The line is the least-squares fit, using $t_i = 5$ min. Note that it does not intercept at zero as required by Eq. (VI-8-1).

(column 5). The deviations of $\tilde{m}$ from m are listed in column 6 and denoted Δ, and the sum of the Δ^2 values (column 7) is 0.42, so the sum of squares s_{yx} is given by

$$(n-2)s_{yx}^2 = 0.42, \qquad \text{i.e.,} \qquad s_{yx}^2 = 0.21$$

The correlation coefficient is 0.9999, so the fit is good (Fig. VI-11a). The lag time is given by the fact that $\ln(m/m_0) = 0$ ($m = m_0$) at this time:

$$t_i = 0.5587/0.11834 = 4.72$$

Hence the disintegration time is not very short, and the only other type of equation that might apply is that of Eq. (VI-8-1). The data are tabulated according to Eq. (VI-8-1) in Table VI-2, and the sum of squares calculated as in the previous case. (Note that the deviations are always taken from

TABLE VI-2

Treatment of Dissolution Data in Example VI-8-1 by Eq. (VI-8-1)

Time (min)	Percent released, $100(m_0 - m)$	$t - t_i$	$\sqrt{t - t_i}$	$100(\tilde{m}_0 - \tilde{m})$	Δ		Δ^2
10	47	5	2.24	49.1	2.1		4.41
15	70	10	3.16	67.4	2.6		6.76
20	83.5	15	3.87	81.5	2.0		4.00
25	91	20	4.47	93.5	2.5		6.25
						Total	10.71

comparable figures; in this case the least-squares fit data are, in each of the two analyses, converted to the same parameter m. It would be erroneous to calculate the sum of squares in the logarithmic form and compare it with the sum of squares in the square root of time form.)

It may be seen that the least-squares fit, using a value of $t_i = 5$ min, is

$$m = 19.93 + 4.41(t - 5)^2$$

with a correlation coefficient of 0.991 and a sum of residuals of

$$s_{yx}^2 = 21.42/2 = 10.71.$$

The fit is therefore less good than for the presentation by Eq. (VI-5-8), because $F = 10.71/0.21 = 51$, which is larger than $F_{crit}(2, 2; p = 0.95) = 19$. To be exact, other values of t_i should be tried and the one giving the smallest value of s_{yx} selected. This has not been done here. The data are shown graphically in Fig. VI-11b.

VI-9 STABILITY OF SOLID DOSAGE FORMS

The decomposition of pure solids has been discussed in Section II-6. However, a solid dosage form is not a pure solid, and several situations exist, all of which are dependent on the "microenvironment" of the solid drug particle. The stability of solid dosage forms, of course, is of great importance, since expiration dating is tied closely to it.

VI-9-1 Pharmaceutical Storage Periods and Expiration Dates

The *life* of a pharmaceutical product is the maximum length of time expected to pass between manufacture and (last) sale. This time is most often controlled by the manufacturer by an expiration period listed on the label. For new products it is customary to calculate the time the product will remain above 90% of label claim (LC). From regression lines of the stability data, here assumed to be of zero order,

$$m = m_0 - Kt \qquad \text{(VI-9-1)}$$

one can calculate the 90% limits ($\pm \gamma$) on (the assay of) m for any extrapolated time $x = t$ (Carstensen and Nelson, 1976):

$$\gamma = t_{n-2,0.1} s_{yx}^2 \left\{ \left[(n+1)/n \right] + (t - \bar{x})^2 / \sum (x - \bar{x})^2 \right\}^{1/2} \qquad \text{(VI-9-2)}$$

where $t_{n-2,0.1}$ is the student t-value, s^2 the sum of squares and m the potency in milligrams per tablet. The potency will in the following always be in this unit. Also, $\bar{x}$ is the average time of the time periods used in the stability study, and n is the number of points. One can now calculate the time t' at which the assay of m will (with 95% confidence) just be at or above 0.9LC (90% of the label claim being a common lower specification limit):

$$m - \gamma = 0.9\text{LC} \quad \text{at} \quad t' \tag{VI-9-3}$$

One refers to t' as the expiration date. Other probabilities than 95% can be used, and other lower limits than 0.9LC can be employed. The factor $(n+1)/n$ is often replaced by $1/n$ (giving the probability of the true mean being above 0.9LC with 95% confidence).

Another method at times used is to calculate the 90% confidence limits on K ($\pm\beta$). If the confidence limits on m_0 are $\pm\alpha$, then

$$0.9\text{LC} = (m_0 - \alpha) - (K + \beta)t' \tag{VI-9-4}$$

Then t' is the (now different) expiration period on this (different) definition.

For products that have been marketed for long periods, the excess $m_0 - \text{LC}$ can be adjusted after scrutinizing stability data. The expiration period in this case should not simply be as long as possible but should be based on a survey of the "market life" of the product. If the number of units produced is N_0 and if at a time t after manufacturing there are N units left unsold, then often

$$N = N_0 \exp(-gt) \tag{VI-9-5}$$

and the time when $N = 0.01N_0$, i.e., when $gt = 4.61$ is the useful market life. If this is less than the calculated value of t', then the economics of shortening the expiration period should be investigated.

VI-9-2 Autodecomposition

It has been stated in Section II-6 that pure solids decompose by sigmoid curves (Fig. II-18). If, in a dosage form, the effect of the microenvironment were nil or negligible, then the stability of the drug would be dictated by the rate with which it, itself, decomposed. This is denoted autodecomposition, and in this case the amount not decomposed (the potency retained) as a function of time would be dependent on time, as shown in Fig. VI-12. Note that in stability work, the potency retained m is usually the parameter of expression. Part AB [Eq. (II-6-14)] is of order zero and may totally dictate the decomposition at room temperature, since, at 25°C, t_B could be a very long time. Part BC is frequently of first order [Eq. (II-6-29)]. Part AB is *truly of zero order* and is written

$$m = m_0 - k_0 t \tag{VI-9-6}$$

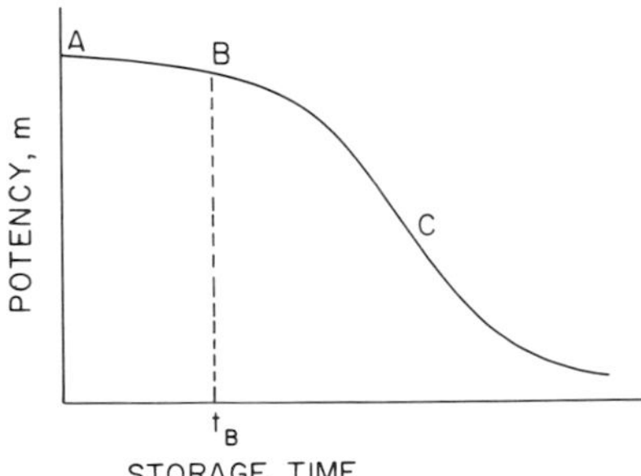

Fig. VI-12 Potency versus time curve for an autodecomposition. Curve AB is of zero order, t_B may be a very long time, e.g., decades, at room temperature, and curve BC is approximately of first order.

where k_0, the zero-order rate constant, in this case is independent of the initial concentration m_0 and the microenvironment.

It has been shown in Section II-6 that topochemical reactions frequently approximate first-order decompositions (Table II-5) and hence such situations (frequently independent of microenvironment) would give rise to first-order decomposition, as would BC in Fig. VI-12. In this case the equation becomes

$$\ln(m/m_0) = \ln[1 - (m_0 - m)/m_0] = -k_1 t \qquad \text{(VI-9-7)}$$

where k_1 is the first-order rate constant and is independent of m_0. From a practical point of view, solid dosage forms are fairly stable, and most often are not considered for marketing if more than a 15% loss occurs, i.e., $(m_0 - m)/m_0 < 0.15$. In this case

$$\ln(m/m_0) = \ln[1 - (m - m_0)/m_0] \sim -(m_0 - m)/m_0 = -k_1 t \qquad \text{(VI-9-8)}$$

hence

$$m = m_0 = k_0' t \qquad \text{(VI-9-9)}$$

where

$$k_0' = m_0 k_1 \qquad \text{(VI-9-10)}$$

Equation (VI-9-9) is *pseudo-zero-order of the first kind* (to distinguish it from other pseudo-zero-order reactions). Note that the pseudo-zero-order rate constant k_0' is dependent on the initial potency [Eq. (VI-9-10)].

VI-9-3 Interactions with Other Substances Other than Water

In most cases the most important decomposition is not the autodecomposition but rather interactions with the microenvironment. These interactions are either with water or with other substances. The latter is often water mediated (i.e., the water serves as a solvent or as a catalyst but does not enter into the reaction scheme, i.e., is not consumed during the decomposition). An example of this is the interaction between phenylephrine (PE) and acetylsalicyclic acid (ASA) (Troup and Mitchner, 1964). Here the reaction (in a simplified fashion) is

$$\mathrm{ASA + PE \rightarrow SA + A(PE)} \qquad \text{(VI-9-11)}$$

i.e., the aspirin is deacetylated to salicyclic acid (SA) and the phenylephrine is acetylated to A(PE). The reaction is more complicated since di- and triacetylated species of phenylephrine also occur. In general one may write

$$\mathrm{A + B \rightarrow C} \qquad \text{(VI-9-12)}$$

where both A and B are somewhat water soluble with solubilities S_A and S_B. The water present, as mentioned in Chapter IV, may be either bound or unbound, and the corresponding volumes are $V = V_b + V_u$. The reaction rate for Eq. (VI-9-12), if the potency of A is followed, may be stated as

$$d[\mathrm{A}]/dt = k_2[\mathrm{A}][\mathrm{B}] = k_2 S_A S_B = k_0'' \qquad \text{(VI-9-13)}$$

Here [A] and [B] are concentrations in the unbound water, which is visualized as a bulk liquid phase of volume V_u. This relates to the total amount of water V by

$$V_u = V - V_b \qquad \text{(VI-9-14)}$$

where V_b is the bound water. If (as is mostly the case) there is more active ingredient A than corresponds to the solubility S_A, then $[A] = S_A$, so that, as implied in Eq. (VI-9-13), at a given temperature, the reaction is *pseudo-zero-order of the second kind* [so called to distinguish it from Eq. (VI-9-10), which derives from the mathematical approximation in Eq. (VI-9-8)]. To convert the concentrations in Eq. (VI-9-13) to amounts retained m, one multiplies by V_u:

$$dm/dt = V_u k_0'' = k_0^* \qquad \text{(VI-9-15)}$$

Note that the pseudo-zero-order rate constant of the second kind k_0^* is proportional to the amount V_u of unbound water but independent of the initial content of drug m_0:

$$m = m_0 - k_0^* t \qquad \text{(VI-9-16)}$$

Example VI-9-1 Aspirin interacts with a drug which is an isoquinoline compound B by a reaction of the type in Eq. (VI-12): A + B = C. In a 500 mg (compression weight) tablet containing 250 mg of aspirin, the zero-order decomposition rate is 1 mg/month at a moisture content of 1% and 1.5 mg/month at 1.2%. What conclusion can be drawn?

Answer The amounts of water are 5 and 6 mg total; i.e., $V_u = 5 - V_b$ and $V_u = 6 - V_b$. Inserting this in Eq. (VI-9-15) gives

$$1.0 = k_0''(5 - V_b) \quad \text{and} \quad 1.5 = k_0''(6 - V_b)$$

Taking the ratio between these gives $1.5 = (6 - V_b)/(5 - V_b)$, from which $V_b = 3$ mg. Hence the unbound water is 2 mg per tablet.

In Eq. (VI-9-12) B can be water. In this case the water content should decrease as the reaction takes place. If there is an abundance of water, so that this decrease does not notably change the number of moles of water present (Leeson and Mattocks, 1958), then the reaction is pseudo-zero-order with a rate constant of

$$k_0^* = Vk_2[H_2O] \qquad \text{(VI-9-17)}$$

Again, this is a function of the moisture content but not of m_0 as long as there is insufficient water to dissolve all the drug.

VI-9-4 Interactions with Moisture

Water is generally not that abundant, and the reaction is

$$\begin{array}{lccc} & \text{A} & +\text{H}_2\text{O} & \rightarrow\text{C} \\ \text{Time zero:} & a_0 & w_0 & 0 \\ \text{Time:} & a - w_0 & 0 & a - w_0 \end{array} \qquad \text{(VI-9-18)}$$

There are a total number of moles a_0 of drug and w_0 of water originally, and when the reaction has gone to completion, there are $a_0 - w_0$ moles of drug left, provided $a_0 > w_0$. The loss in such a dosage form therefore depends on the amount of water present (Carstensen *et al.*, 1964, 1966). The time profile will be as shown in Fig. VI-13a. This type of reaction is characterized by (a) distinctly different loss rates in different batches (since different batches will have different moisture contents) and (b) the same ultimate loss (in one batch) regardless of temperature (since the amount of moisture is not a function of the temperature). This last point requires some comment regarding containers.

For accelerated testing *the closure must be hermetic*, otherwise the preparation dries out at the higher temperatures, and can for instance appear to be more stable at 55°C than at 45°C. The best means of a hermetic closure is to seal the preparation in a tube by glass blowing. Next best is using a flowed-in rubber gasket in the lid or dipping the lid of the closed bottle in molten carnauba wax. This latter has a high melting point and solidifies rapidly on cooling and does not melt in the usual storage conditions (below 60°C).

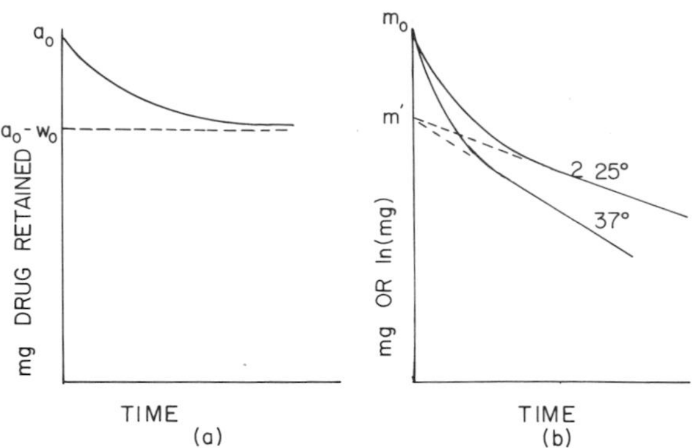

Fig. VI-13 (a) Decomposition profile for an interaction reaction (A + B → C). (b) Decomposition profile for an interaction reaction (A + B → C) and an autodecomposition (A → D) occurring simultaneously.

The volume of the containers should be noted (V_c) and the amount n (moles) of moisture evaporated into the headspace at the temperature [$T(°\mathrm{K}) = T(°\mathrm{C}) + 273$] of storage calculated.

$$V_c(P_{H_2O} - 24.5)/760 = n83T \tag{VI-9-19}$$

Here P_{H_2O} is water vapor pressure at T(°K). The loss is usually insignificant compared with the total amount of moisture in *all* the tablets in the container.

If an autodecomposition or interaction occurs simultaneously with the hydrolysis, then the time profile will be as shown in Fig. VI-13b. Note that the linear extrapolation (whether to zero or to first order) intersects the ordinate at the same point (mass m' of drug retained). If the molecular weight of the drug is M, then

$$w_0/18 = (m_0 - m')/M \tag{VI-9-20}$$

represents the number of moles of unbound water in the preparation.

Example VI-9-2 A drug of molecular weight 183 is present in an amount of 100 mg of drug per tablet and exhibits the following decomposition data at 25°C: 10% lost after 0.5 year, 13% lost after 1 year, and 19% lost after 2 years. What would be an estimated figure for the unbound moisture?

Answer The decomposition is of zero order and the equation used is

$$m = 93 - 6t \tag{VI-9-21}$$

where t is time in years. The amount of unbound moisture is therefore

$$w_0 = 18(100 - 93)/183 = 0.69 \quad \mathrm{mg}$$

VI-9-5 Accelerated Testing

The principles of accelerated testing (e.g., Arrhenius plotting) *can* be used (with caution) for solid dosage forms, but *the actual rate limiting process must first be known*. Casual data treatment can give rise to severe miscalculations. Several advantages exist in the accelerated testing: It allows (a) estimation of the room temperature rate constant rather rapidly (by extrapolation), (b) determination of the reaction order (which facilitates later actual treatment of long-term room temperature data), and (c) estimation of stability at other (and fluctuating) temperature situations than constant room temperature (so-called temperature cycling).

In extrapolating to room temperature a statistical problem exists because ordinarily the rate constants k are calculated as the slopes from n time points at p different temperatures, and k at 25°C then calculated from inserting $1/T = 0.003355$ (corresponding to 25°C) in the equation

$$\ln k = (-E/R)(1/T) + B \qquad \text{(VI-9-22)}$$

The limits on this k value can be calculated from Eq. (VI-9-2), where $x = 1/T$, $t = 0.003355$ (and m corresponds to $\ln k$). The number of degrees of freedom is $p - 2$, and the t-value is $t_{2,0.05} = 4.3$, i.e., rather large. This is in spite of the fact that a total of 12 assays (plus one original) have been performed. Carstensen and Su (1971) have treated this problem by pointing out that what is actually sought is the extrapolation of the potency m at 25°C to a certain time period (e.g., 36 months), and that therefore the equation to study is

$$m = m_0 \exp(-kt) = C m_0 \exp[-t \exp(-E/RT)] \qquad \text{(VI-9-23)}$$

This equation has three unknown parameters, and if the n points at each of the p temperatures are used individually, then the number of degrees of freedom is $np - 3$ and the student t value is then much smaller. If three time points were used in the above example, then $df = 4 \times 3 - 3 = 9$, and $t_{9,0.05} = 2.26$ i.e., much less than the value of 4.3 quoted above.

The method of Carstensen and Su (1971) uses the Gauss–Newton approximation and is complicated. Carstensen (1977) and Slater *et al.* (1979) have suggested an approximate method using individually calculated k values from each point at each temperature, i.e., by this means also having $np - 3$ degrees of freedom. Slater *et al.* (1979) used this method to predict adequately the stability of vitamin A in a tablet dosage form.

Certain *errors* are frequently committed in accelerated studies of the solid dosage form, notably the following:

(a) One-point kinetics are used (i.e., the order of the reaction remains unknown).

(b) Containers that are not tight are used.

(c) Initial assays are performed by a control assay and the remainder by a (different) stability indicating assay.

(d) Certain aspects of the stability profile are neglected; it often happens, for instance, that the first time points after the initial assay (e.g., the 6 month point at room temperature) give higher assay results than initially. This can be due to a decomposition product coupled with an assay method which is not sufficiently specific.

(e) Temperature is poorly controlled and not uniform in the constant temperature ovens (stations).

Certain *pitfalls* exist in accelerated testing:

(a) There may be competing reactions where extrapolation will give false "lows" because it will extrapolate the reaction which is predominant at the higher temperature. At times this is different from the reaction predominant at room temperature (for sulfanilamides, for example).

(b) There may be decomposition of excipients. The decomposition products could be more or less reactive than the excipient itself.

(c) The physical nature of the drug (or an excipient) can change. For example, the drug may melt and the liquified drug then has a much greater possibility of interaction with excipients.

(d) The drug may vaporize (e.g., nitroglycerin).

REFERENCES

Bamba, M., Puisieux, F., Marty, J.-P., and Carstensen, J. T. (1979). *Int. J. Pharmaceutics* **3**, 87.

Bathe, R. V., Häffliger, O., Langenbucher, F., and Schönleber, D. (1975). *Pharm. Acta Helv.* **50**, 3.

Berry, H., and Ridout, C. W., (1950). *J. Pharm. Pharmacol.* **2**, 619.

Carstensen, J. T. (1973). "Theory of Pharmaceutical Systems," Vol. II, pp. 287–288. Academic Press, New York.

Carstensen, J. T. (1974). *In* "Dissolution Technology" (L. Leeson and J. T. Carstensen, eds.), pp. 1–20. Industrial Pharmacy and Technology Section of Acad. Pharmaceutical Sci., Washington, D.C.

Carstensen, J. T. (1976). Abstracts, Amer. Pharmaceutical Assoc./Acad. Pharmaceutical Sci. National Meeting **21** (Abstracts, **6** (2), paper 79).

Carstensen, J. T. (1977a). "Pharmaceutics of Solids and Solid Dosage Forms," pp. 64–68, 76, 106. Wiley, New York.

Carstensen, J. T. (1977b). How long and at what risk. Paper presented at *Ann. Conf. Pharmaceutical Analysis, 17th, Lake Delton, Wisconsin*, University of Wisconsin Extension, Madison.

Carstensen, J. T., and Dhupar, K. (1976). *J. Pharm. Sci.* **65**, 1634.

Carstensen, J. T., and Nelson, E. (1976). *J. Pharm. Sci.* **65**, 311.

Carstensen, J. T., and Su, K. (1971). *Bull. Parenter. Drug Assoc.* **25**, 287.
Carstensen, J. T., Johnson, J. J., Valentine, W., and Vance, J. J. (1964). *J. Pharm. Sci.* **53**, 1050.
Carstensen, J. T., Aron, E., Spear, D., and Vance, J. J. (1966). *J. Pharm. Sci.* **55**, 561.
Carstensen, J. T., Wright, J. L., Blessel, K. W., and Sheridan, J. (1978a). *J. Pharm. Sci.* **67**, 48, 982.
Carstensen, J. T., Lai, T. Y. -F., and Prasad, V. K. (1978b). *J. Pharm. Sci.* **67**, 1303.
Couvreur, P. (1975). Thesis, Docteur en Sciences Pharmaceutiques, p. 87. Univ. Catholique de Louvain, Belgium.
Fessi, H., Marty, J. -P., Puisieux, F., and Carstensen, J. T. (1979). *Int. J. Pharmaceutics* **1**, 265.
Fessi, H., Marty, J. -P., Puisieux, F., and Carstensen, J. T. (1980). *J. Pharm. Sci.* (in press).
Higuchi, T. (1963). *J. Pharm. Sci.* **52**, 1145.
Ingram, J. T., and Lowenthal, W. (1966). *J. Pharm. Sci.* **55**, 614.
Ingram, J. T., and Lowenthal, W. (1968). *J. Pharm. Sci.* **57**, 393.
Kennon, L., and Swintosky, J. V. (1958). *J. Am. Pharm. Assoc. Sci. Ed.* **47**, 397.
Khan, K. A., and Rhodes, C. T. (1975). *J. Pharm. Sci.* **64**, 166.
Khan, K. A., and Rhodes, C. T. (1976). *J. Pharm. Sci.* **65**, 1837.
Kitamori, N., and Iga, K. (1978). *J. Pharm. Sci.* **67**, 1436.
Kitamori, N., and Shimamoto, T. (1976). *Chem. Pharm. Bull.* **24**, 1789.
Kohler, H. J., Soliva, M., and Speiser, P. (1975). *Pharm. Acta Helv.* **50**, 17.
Langenbucher, F. (1969). *J. Pharm. Sci.* **58**, 1265.
Langenbucher, F. (1972). *J. Pharm. Pharmacol.* **24**, 979.
Leeson, L., and Mattocks, A. (1958). *J. Am. Pharm. Assoc. Sci. Ed.* **47**, 329.
Levich, V. G. (1962). "Physicochemical Hydrodynamics," p. 342. Prentice-Hall, Englewood Cliffs, New Jersey.
Lippold, B. C. (1977). *In* "Formulation and Preparation of Dosage Forms" (J. Polderman, ed.), p. 215. Elsevier/North Holland, Amsterdam, Holland.
Nelson, K. G., and Wang, L. Y. (1977). *J. Pharm. Sci.* **66**, 1758.
Nelson, K. G., and Wang, L. Y. (1978). *J. Pharm. Sci.* **67**, 87.
Nogami, H., Nagai, T., and Uchida, H. (1966). *Chem. Pharm. Bull.* **14**, 152.
Prescott, L. F., Steel, R. F., and Ferrier, W. R. (1970). *Clin. Pharmacol. Ther.* **11**, 496.
Slater, J. G., Stone, H. A., Palermo, B. T., and Duvall, R. N. (1979). *J. Pharm. Sci.* **68**, 49.
Takieddin, M., Puisieux, F., Didry, J. R., Touré, P., and Duchenê, D. (1977). *Int. Conf. Powder Technology, 1st, Paris, May 31, 1977*. Abstracts, Vol. I, p. 248.
Theeuwus, F. (1975). *J. Pharm. Sci.* **64**, 1987.
Tingstad, J. E., and Riegelman, S. (1970). *J. Pharm. Sci.* **59**, 692.
Troup, A., and Mitchner, H. (1964). *J. Pharm. Sci.* **53**, 375.
Wagner, V. (1969). *J. Pharm. Sci.* **58**, 1253.
Washburn, E. H. (1921). *Phys. Rev.* **17**, 273.

Appendix

A-I *Prefixes for Multiples and Fractions*

Prefix	Symbol	Factor
tera	T	10^{12}
giga	G	10^{9}
mega	M	10^{6}
kilo	k	10^{3}
deci	d	10^{-1}
centi	c	10^{-2}
milli	m	10^{-3}
micro	μ	10^{-6}
nano	n	10^{-9}
pico	p	10^{-12}

A-II *Conversion Factors for SI Units (Système International)*

Dimension	Unit	Symbol	Definition	Common conversions
Length	Meter	m		1 in. = 0.0254 m 1 ft = 0.3 m 1 Å = 10^{-10} m
Mass	Kilogram	kg		1 lb = 0.4536 kg
Time	Second	s, sec		
Temperature	Kelvin	K		°F unit = (1/1.8) K unit °C unit = K unit Zero points: 0°C = 273.15 K = 32°F
Area	Square meter	m^2		1 in.^{2} = 6.45 10^{-4} m^2 1 ft^2 = 0.0929 m^2
Volume	Cubic meter	m^3		1 in.^{3} = 1.639 10^{-5} m^3 1 gal = 3.785 10^{-3} m^3 1 liter = 10^{-3} m^3
Density	Kilogram/ cubic meter	kg/m^3		1 g/cm^3 = 10^3 kg/m^3 1 lb/in.^{3} = 2.768 10^{-3} kg/m^3
Energy	Joule	J	kg m^2/sec^2	1 cal = 4.1868 J 1 BTU = 1055 J
Force	Newton	N	kg m^{-1} sec^{-2} = J m^{-1}	1 kg force = 9.80 N 1 lb force = 4.48 N
Pressure	Pascal	Pa	kg $m^{-1}sec^{-2}$ = N m^{-2}	1 atm = 101 kPa = 101 kN m^{-2} lb/in.^{2} = psi = 6895 Pa = 6.895 kN m^{-2}

Author Index

Numbers in italics refer to the pages on which the complete references are listed.

C

D

E

F

G

U

V

W

Y

Z

Subject Index

W

Y